Praise for Math Puzzles and Brainteasers

Terry Stickels combines his masterful ability to create diverse, challenging and just plain fun puzzles with a wide range of math concepts, in a playful way that encourages the solver to discover their own unique methods of finding solutions.

—**David Kalvitis,** author of *The Greatest Dot-to-Dot Books in the World*

Logical, numerical, visual-spatial, and creative thinking problems can all be found within these covers, embracing a wide spectrum of thinking skills for developing minds. Terry Stickels also encourages indulgence in mathematical play, which for young students is an indispensable component of motivated and successful problem solving.

—**Barry R. Clarke,** *Mind Gym* compiler, *The Daily Telegraph* (UK)

Even kids who are not math nerds will enjoy this book. Stickels hits the perfect mix of brainteasers: They're challenging while still managing to be great fun at the same time!

—**Casey Shaw,** Creative Director, *USA WEEKEND* magazine

Terry Stickels is clearly this country's Puzzle Laureate. He has concocted a delightful and challenging volume of brainteasers that belong in every math teacher's library. Focused specifically on grades 3–5 and grades 6–8, these puzzles both educate and sharpen children's critical thinking skills. As an award-winning puzzle constructor myself, I am always in awe of what Terry comes up with.

—**Sam Bellotto Jr.,** Crossdown

Jossey-Bass Teacher

Jossey-Bass Teacher provides educators with practical knowledge and tools to create a positive and lifelong impact on student learning. We offer classroom-tested and research-based teaching resources for a variety of grade levels and subject areas. Whether you are an aspiring, new, or veteran teacher, we want to help you make every teaching day your best.

From ready-to-use classroom activities to the latest teaching framework, our value-packed books provide insightful, practical, and comprehensive materials on the topics that matter most to K–12 teachers. We hope to become your trusted source for the best ideas from the most experienced and respected experts in the field.

MATH PUZZLES
and
BRAINTEASERS,
Grades 6–8

Over 300 Puzzles that Teach
Math and Problem-Solving Skills

Terry Stickels

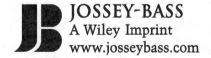
JOSSEY-BASS
A Wiley Imprint
www.josseybass.com

Contents

Contents

Foreword

In the 1950s we punished misbehaving students in middle school
or junior high school by making them stay after school to do 100
multiplication or long-division math problems. Mathematics was
taught in a highly ordered and mechanically repetitive manner
with the objective being to master basic arithmetic skills. A decade
later New Math emerged as a response to the Soviet Union's early
dominance in the race to outer space. The focus in the classroom
shifted from drill to conceptual understanding to prepare students
for early exposure to advanced mathematics. This theoretical
approach failed miserably because middle school students entering
high school did not possess the arithmetic skills necessary to do
calculations in mathematics and science courses.

Consequently, the 1970s saw a backlash and the Back-to-the-Basics
movement emerged. During the last three decades we have witnessed
the emergence of several more movements in mathematics education
attempting to balance the mastery of computational skills and
conceptual understanding. These movements included a problem-
solving approach, a high-tech calculator/computer approach, and a
mathematical user-friendly approach. Why have all these movements
fallen short? The number-one complaint about mathematics from
students is: Math is boring! Perhaps, in the 21st century there should
be a Math for Fun movement that makes mathematics exciting,
challenging, and rewarding. With this puzzle book, Terry Stickels

has pioneered an innovative approach to motivating students not only to learn and do mathematics but also to come away from the experience with a positive attitude.

In this creative and stimulating book, students and teachers will find a potpourri of mind-expanding puzzles designed to enhance and enlighten, as well as to entertain. This book contains an incredible assortment of puzzles of various types including logical, geometrical, mathematical, and verbal. The diversity of the puzzles and the various approaches to solving them will improve the student's problem-solving skills, as well as the general thinking skills required for subjects outside of mathematics.

Using these puzzles as supplements to the traditional mathematics curriculum, the teacher can add a new dimension to students' learning experience. For example, the puzzles can be used to introduce and motivate fundamental algebraic or geometrical concepts. The puzzles also can be used to apply these abstract concepts to concrete problems. Thus, these puzzles can supplement the traditional "story problems" that have been tormenting students for decades. In addition, the puzzles can be used to give bonus points or extra credit to students who finish their required daily assignments in a timely manner. An exciting and promising application of the puzzles would be in a math club where the students can compete and play games that challenge the mind and the creative spirit.

Teaching and learning mathematics constitute a multidimensional experience. The first dimension consists of the rules and algorithms required to do calculations. Many students perceive mathematics as being one-dimensional—"number crunching." To apply the potential power of mathematics effectively, however, they need a second dimension: a conceptual understanding and framework. To expand the potential power of mathematics, they must enter a third dimension: the intuitive and mind-expanding creative process. In this dimension we are "thinking outside the box." This puzzle book by Terry Stickels will take students and teachers on a multidimensional journey filled with multilevel thinking, creative and imaginative explorations, and exciting discoveries and solutions.

One of the great challenges of teaching mathematics in the 21st century is how to reach the diverse student population and how to teach to a variety of individual learning styles. Some students are visual learners; others learn by studying concrete examples, and many students with strong verbal skills learn by translating the mathematical symbolism to words. This puzzle book offers the potential to be an effective alternative approach to solving this universal challenge. The flexibility and variety of these puzzles that span the whole spectrum of traditional mathematics in Grades 6–8, and the direct application of these puzzles to a wide range of learning styles, will make this Math for Fun approach a rewarding and positive experience for students and teachers alike.

In conclusion, this puzzle book can be used to reward the motivated hard-working students but also to "punish" the misbehaving students by making them stay after school to do 100 Terry Stickels puzzles! Serendipitously, Terry Stickels has created an innovative way to reach out to troubled students that can change their negative attitudes to positive experiences and a positive perspective on thinking and creativity.

February 2009

Dr. John Konvalina
Department of Mathematics
University of Nebraska at Omaha

Acknowledgments

This book would not have been possible without the work and suggestions of the following people:

Mr. Sam Bellotto Jr. of *CROSSDOWN.COM*

Ms. Terry Baughan of *TALLROSE PRODUCTIONS*

Ms. Shelley Hazard of *PUZZLERSPARADISE.COM*

Mr. Barry Finnen of *PHYSICS247.COM*

Webmaster Mr. Roger Smith

Mr. Robert Webb of *SOFTWARE3D.COM*

Ms. Suzanne Alejandre of *THE MATH FORUM@DREXEL*

Mr. Martin Gardner

Mr. Casey Shaw of *USA WEEKEND* magazine

Mr. Brendan Burford of *KING FEATURES*

Ms. Kelsey Flower

Mr. Alex Stickels

Finally—a special thanks to my right hand and the person who makes all this happen, Ms. Christy Davis, owner of Executive Services, Arlington, Texas.

About This Book

*A good math puzzle, paradox, or magic trick can stimulate a
child's imagination much faster than a practical application . . .
and if the game is chosen carefully, it can lead almost effortlessly
into significant mathematical ideas.*

Martin Gardner
America's Mathemagician

Mr. Gardner's quote captures one of the main reasons for this book.
My intention was (and has been with all my books) twofold: to
provide challenging fun and to offer options to think differently—
and maybe discover opportunities to become a better thinker.

There are countless stories of great thinkers being puzzle-lovers, but
have you ever wondered why that is so? What is the connection
between creative, bright people and their insatiable thirst for puzzles?

The father of modern-day puzzle writers, Henry Dudeney, gave us
one clue when he said, "Puzzles, like virtue, are their own reward."
He also noted that "the fact is that our lives are largely spent solving
puzzles; for what is a puzzle but a perplexing question? And from our
childhood upwards we are perpetually asking questions or trying to
answer them."

A well-crafted puzzle seems to naturally encourage a nontraditional or more circuitous route to its solution. This emphasis on different approaches is like brain candy for thinkers. Thinkers begin by dissecting a puzzle and viewing it from different perspectives simultaneously. Great puzzle solvers enjoy twisting, bending, separating, and spinning a puzzle. They look at it backward, forward, upside down, and sideways. Is there a quick solution? More than one solution? What kinds of internal patterns do puzzles have? Can I get to the answer and then make it into a new puzzle? Does it have direct application to the real world?

A quick comment about being a "world-class" thinker: You don't have to be a genius to be one. You may have noticed with many of the stories about successful, bright people that they often are accompanied by a back story about how they weren't the best students in class or weren't initially successful in certain studies. Einstein and Edison come to mind. Maybe they already were employing their own ideas of arriving at solutions that weren't considered acceptable practices at the time. Students are often inspired by these stories not to give up working to become better thinkers because they don't fit mainstream profiles.

Here's one other important point, often lost in this day and age of timed standardized tests: The time it takes to solve any problem or puzzle has nothing to do with mental ability or intellectual level. Some of the greatest thinkers deliberately take their time—enjoying the play of a puzzle and savoring it like a good meal!

The puzzles in this book are designed to sharpen the creativity and problem-solving skills, as well as the mathematics content skills, of students in grades 6–8.

This book is designed with the following objectives:

- Offer a panoramic approach to the thinking skills that kids need to excel in math

- Incorporate a broad spectrum of different kinds of puzzles
- Meet the grade-appropriate guidelines set forth by the National Council of Teachers of Mathematics
- Venture into content areas where previous math and thinking skills books have not gone
- Be challenging, but also offer lots of fun along the way

Although the puzzles are easy, medium, and difficult, none are so designated. What one student will find easy, another may see as difficult, and vice versa. A difficulty rating also might be intimidating to some students, and interpreted as a good reason for *not* solving a puzzle—the opposite of the book's purpose.

The range of puzzles incorporates multiple approaches to skill-building, including numerical manipulation, spatial and visual problems, and language arts exercises. There is no one "best" pathway to solve each puzzle, and often numerous entry points to find solutions. Students invariably will find the way, using a mix of intuition and thinking skills that are uniquely their own.

Puzzles can offer an experience parallel to a new dive off a diving board—an exciting intuitive leap into the unknown, with possible scintillating results! If you look at some of the greatest discoveries of science and mathematics, they often are accompanied by intuitive leaps, supported in turn by clear thinking, logic, evidence, and repeated consistent results in trials.

In this complex world, we need all the good thinkers we can get—and grow. Any opportunity we have to help uncover those talented young people should be welcomed and maximized. Fortunately for me, I live in a time when puzzles are being used increasingly in schools and businesses to promote critical thinking and mental flexibility. A stream of recent studies makes clear that puzzles and games indeed will help the capacity of your mind to grow and stay flexible. These puzzles also can contribute to the development of an expanding cadre of lifelong learners—a primary goal of every parent and educator.

The Author

Terry Stickels is dedicated to helping people improve their mental flexibility and creative problem-solving capabilities through puzzles— and making it fun. His books, calendars, card decks, and newspaper columns are filled with clever and challenging exercises that stretch the minds of even the best thinkers. And he especially enjoys creating puzzles for kids.

Terry is well known for his internationally syndicated columns. *FRAME GAMES,* appearing in *USA WEEKEND* magazine, is read by more than 48 million people in six hundred newspapers weekly. *STICKELERS,* published daily by King Features, appears in several of the largest newspapers in America, such as the *Washington Post,* the *Chicago Sun-Times,* and the *Seattle Post-Intelligencer.* Terry also is the featured puzzle columnist for *The Guardian* in London—the United Kingdom's largest newspaper.

As a highly popular public speaker, Terry's keynote addresses are fast-paced, humorous looks at the ability (and sometimes the lack thereof) to think clearly. Distinguished authorities such as the National Council of Teachers of Mathematics also praise his work as an important aid in assisting students to learn how to think critically and sharpen their problem-solving skills.

Born and raised in Omaha, Nebraska, Terry was given his first puzzle book at age eleven. Fascinated by the book's mind-bending playfulness, he soon was inventing puzzles on his own—lots of them. He attended the University of Nebraska at Omaha on a football scholarship. While he was at UNO tutoring students in math and physics, he saw the advantages of using puzzles to turbocharge understanding of several concepts within those disciplines.

After several years as an occasionally published creator of puzzles, Terry was asked to produce a weekly column for a twelve-newspaper syndicate in Rochester, New York. Two years later his puzzles caught the attention of Sterling Publishing in New York. His first book, *MINDSTRETCHING PUZZLES*, became an immediate hit and continues to sell well to this day. Twenty-five more puzzle books have followed, three of them sponsored by the high-IQ society Mensa.

Terry lives in Fort Worth, Texas, where he is working on his next generation of puzzles to once again captivate, challenge, and delight his worldwide readership.

Introduction

I love puzzles. One guy tries to make something to keep another guy out; there must be a way to beat it.

Richard Feynman
Nobel Laureate—Physics

This book contains more than 300 puzzles, ranging from relatively easy word puzzles to more difficult math brainteasers and requiring math skills from addition and subtraction to determining probability and algebraic thinking. Here are the types of puzzles you will find within these pages:

Mathematical	Frame Games
Spatial/Visual	Cryptograms
Logical	Analogies
Analytical Reasoning	Sequence
Word puzzles	Sudoku

By design, I have included a large number and broad spectrum of puzzles, providing teachers and students with multiple options. The puzzles are organized into chapters on numbers and operations; geometry and measurement; mathematical reasoning; and algebra, statistics, and probability, to facilitate the instructor's ability to enhance areas of the curriculum that are most appropriate for their application, adding richness, change of pace, and reinforcement to the teaching and learning processes.

Some Puzzle-Solving Tips

Puzzle solving is sometimes like mathematical problem-solving, but at other times you have to move away from standard approaches to learning mathematics when solving these puzzles. Think about the puzzles from different perspectives and with a sense of play. Consider some of the following:

- Can the puzzle be solved by breaking it down into simpler components?

- Do any patterns repeat often enough to suggest a prediction for "what comes next"?

- Does the puzzle have multiple answers, or at least one optional answer?

- Try thinking of ways to twist, bend, separate, or spin the puzzle. What does it look like backward, forward, upside down, and sideways?

- Does your answer make sense? Can you plug your answer back into the question to check all the parameters?

- If your answer seems absurd or counterintuitive, can you still defend it? Your answer may very well be correct, even though it seems strange or unusual.

- Don't worry about how you might be seen if you can't solve the puzzle. We all make mistakes, and no one can answer every question. Just relax, have a good time, and don't worry about other people's opinions.

Projects throughout the book marked with a ✍ symbol can be done using easy-to-find manipulatives, such as coins, blocks, and cut paper, to help students who may have trouble visualizing some of the puzzles.

You may wonder why some language arts puzzles are included in a math puzzle book. Puzzles and problems such as analogies and analytical reasoning, which are more "language arts" in nature, promote and augment critical thinking skills. Take the *FRAME GAMES*, for example. *FRAME GAMES* are words, letters, pictures, fonts, and the like, juxtaposed in a way to reveal a common idiom, famous person, athlete, movie, song title, and so on. These include components of spatial/visual thinking, language, memory, vocabulary, and light-hearted fun. When people solve even one puzzle correctly—and find the fun in doing so—they are eager to jump to the next challenge, even if it's a puzzle of a different kind. It has also been found that solving a type of puzzle in one area often triggers the mind into a flexible mode that makes it easier to solve problems and puzzles in other areas.

Another appealing feature of the *FRAME GAMES* is that they don't always follow the standard left-to-right or top-to-bottom pattern for their solutions. Mental flexibility from different perspectives is required. These puzzles can be used in a broad spectrum of classroom situations—from special education to warm-ups in calculus classes. You'll find them placed strategically throughout the book. They offer both a kind of mental break and a different type of thinking challenge.

There is no wrong way to use these puzzles. They are meant to be treated like a good watch or pair of shoes—to be used over and over again. And they never wear out!

Some application ideas are:

- As warm-ups to introduce a new element of math curriculum

- As a focus for competition among teams

- To inspire students to create their own versions to share with their classmates

- As a feature on posters or in class newsletters as the puzzle of the day, week, month, or holiday

- For group problem-solving

- To be sent home for sharing with friends and family

- As the basis for discussions on how the concepts of certain puzzles might have real-life applications and how they might be used within different professions

- Chosen randomly, for the sole purpose of personal entertainment

I hope the solvers of these puzzles will see that a book of puzzles can be much more than just a puzzle book. It is also a wonderful way to be involved in creative and challenging "mindstretch" exercises that will enhance a lifetime of thinking skills in so many ways.

So read on . . . and enjoy!

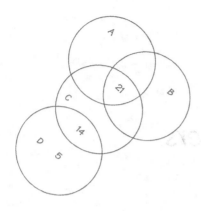

Part I

BASE
+ BALL
GAMES

NUMBERS
and
OPERATIONS

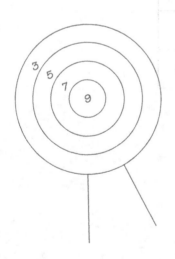

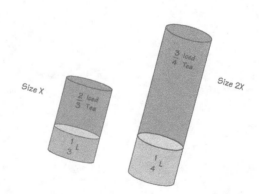

Size X

Size 2X

Whole Numbers

1. Below is a partial Magic Square using the numbers 1–16. The rows, columns, and diagonals must each total the same sum. Place the final four numbers in the appropriate squares.

16	9	2	7
6	?	?	13
11	?	?	4
1	8	15	10

2. Try your luck at these three analogy puzzles. An analogy compares two things to two others. Here is an example:

$$9 : 10 :: 81 : ?$$

This is read as "9 is to 10 as 81 is to . . ." and the answer is 100.

$$9^2 = 81$$

$$10^2 = 100$$

Now find the missing numbers or words in the following analogies:

a. $11 : 121 :: 20 : ?$

b. $3 : 27 :: 4 : ?$

c. Hexagon : 6 sides :: ? : 7 sides (Possible answers: Nonogon, Sevegon, Hedagon, Heptagon)

3. What is the missing number in the sequence below?

| 13 | 57 | 91 | 11 | 31 | 51 | _?_ |

HINT
Don't be afraid to move these numbers around and look at it from another perspective.

4. The numbers in the center of the pyramids are related to the numbers at each corner. What is the missing number in the middle of the last pyramid?

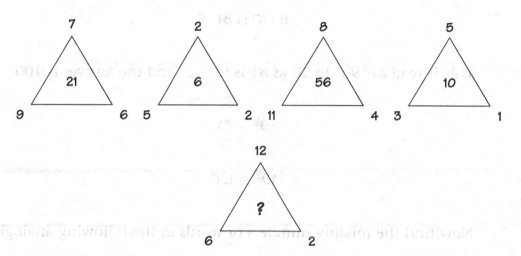

5. Alphametics are puzzles where each letter stands for a different digit. Here are two alphametics. No word can begin with a zero. We've started you out with some of the letters.

a.
```
   BASE
 + BALL
  GAMES
```
Let E = 3
B = 7
A = 4

b.
```
   MAD
    AS
  +  A
  BULL
```
Let D = 3
L = 7
A = 8

6. Tom walks half a distance at 4 mph and runs back the other half at 12 mph. What is his average speed for the entire trip?

7. If 32 people were to enter a statewide singles tennis tournament, how many matches would be played, including the championship?

8. Below is a sequence where the difference between each successive pair of numbers is 6. What is the 500th number in this sequence?

5   11   17   23   29   35   41   . . .

> *HINT*
> *Make a chart to help you figure out the pattern, then apply the pattern to the 500th number.*

9. The seven sets of numbers below all have a certain logic that is the same in all seven numbers. See if you can determine the relationship and come up with the final digit of the last number.

a. 1 3 8 2 1   e. 2 5 1 1 6

b. 5 1 3 4 2   f. 4 4 4 0 3

c. 6 0 0 2 7   g. 7 0 3 1 <u>?</u>

d. 9 2 0 4 0

10. In a laboratory, two sub-atomic particles are being crashed into each other as part of an experiment. Particle #1 is moving directly toward particle #2 at 15 mers a second. Particle #2 is moving directly toward particle #1 at a speed of 25 mers a second. The particles are 1,000 mers apart when they begin moving. What will be the distance between them 1 second before they crash?

> *HINT*
> *Don't be concerned about the term "mers." It's a fictitious distance.*

Just for Fun: Frame Game

11. Find the hidden word or phrase.

FRAME

Laugh, laugh, laugh,
laugh, laugh, laugh, laugh,
laugh, laugh, laugh, laugh,
laugh, laugh, laugh, bank

© 2009 Terry Stickels

GAMES

12. If one can of dog food feeds eight pups or six dogs, then eight cans of the same dog food will feed 40 puppies and how many dogs? Be careful with the puzzle. It is asking for puppies *and* dogs, not puppies *or* dogs.

13. Inserting the numbers 4, 3, and 9 once and only once, and any of the four operations (+, −, ×, ÷) in the grid below, see if you can total the numbers represented in each row and column.

4	×	3	−	9	+	8	11
+	▓	+	▓		▓		
8		9	+	3		4	71
+	▓		▓	+	▓	−	
9				8		3	41
×	▓		▓		▓		
	+	8	+	4			47

39   31   38   38

HINT
Remember the correct order of operations.

14. The grid below has symbols that contain a whole number value less than 10. Each symbol has its own value. The numbers you see at the end of each row and column are the sums of the figures' values for that row or column. Can you find out the value of each symbol?

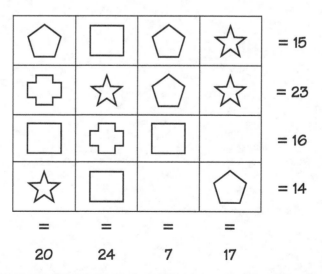

15. The numbers in the box go together in a certain way. See if you can determine the pattern and come up with the two missing numbers.

7	42	6
6	30	5
4	12	3
5	?	?

HINT
There are two different relationships to work out here.

16. What number comes first in the sequence below?

?   1   5   4   8   7   11   10   14   13   17

17. The boxes below have a certain logic that enables you to predict what the two missing numbers are. Can you find them?

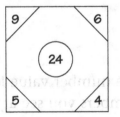  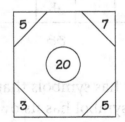 

18. What is the missing number?

5   25   125   _____   3,125   15,625

19. One of the numbers below does not belong with the others for a simple, straightforward reason. Which is the odd one out?

13,754

14,933

16,283

16,637

16,175

18,911

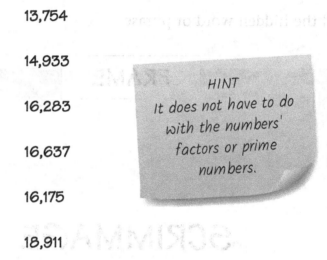

HINT
It does not have to do with the numbers' factors or prime numbers.

20. How many zeros does a thousand thousand million have?

Just for Fun: Frame Game

21. Find the hidden word or phrase.

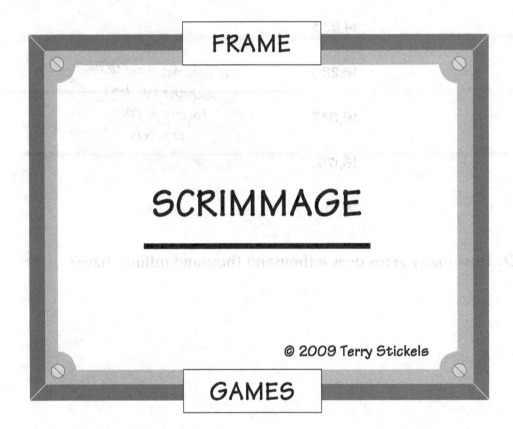

FRAME

SCRIMMAGE

© 2009 Terry Stickels

GAMES

22. On the planet Leptron, there are two types of people—Crizellas and Frizellas. Crizellas have 4 heads and Frizellas have 11 heads. Both sets of people all look identical.

When I last visited them, my friend Arzella said, "I see 53 heads in the Trizella swimming pool, so I know exactly how many Crizellas and Frizellas are in the pool."

Now, can you tell me how many of each are in the swimming pool?

23. Our number system is called $Base_{10}$ because we start to repeat the sequencing of numbers after the number 9. It is possible to have any number system, even systems based on fractions.

Here is what the first nine numbers in $Base_9$ look like:

1 2 3 4 5 6 7 8 10

The number 9 in $Base_{10}$ becomes the number 10 in $Base_9$. How would you write the number 22 in $Base_{10}$ in the $Base_9$ system?

24. If you write down all the numbers from 1 to 100, how many total single digits will you have written down? Include both 1 and 100.

25. Below is a number pyramid where the numbers are arranged in a logical fashion so you can replace each question mark with a correct whole number. Determine what that logic is and find each missing number.

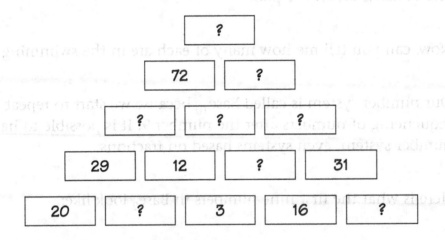

26. Larry and Jane went to play Bingo for a fundraising event. The two got very excited as the night progressed, particularly when Larry announced that he had a BINGO! Below are three statements that he made to Jane. Two of the statements are true and one is false. Can you determine which of his statements was false and where his BINGO was?

B	I	N	G	O
8	16	32	60	72
12	20	34	53	69
13	30	Free	48	64
7	27	44	48	61
2	19	39	51	74

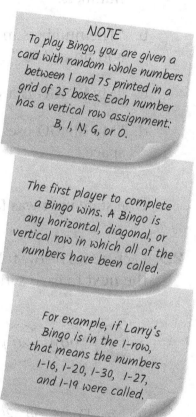

NOTE
To play Bingo, you are given a card with random whole numbers between 1 and 75 printed in a grid of 25 boxes. Each number has a vertical row assignment: B, I, N, G, or O.

The first player to complete a Bingo wins. A Bingo is any horizontal, diagonal, or vertical row in which all of the numbers have been called.

For example, if Larry's Bingo is in the I-row, that means the numbers 1-16, 1-20, 1-30, 1-27, and 1-19 were called.

Larry said:

a. "My BINGO has all even numbers."

b. "My BINGO has two numbers evenly divisible by 5."

c. "All of the numbers in my BINGO are evenly divisible by 4."

27. A little while after Larry's BINGO, Jane got her own BINGO. Her face flushed with excitement, she yelled out "BINGO!" With a smirk, she turned to Larry and said:

a. "My BINGO has all prime numbers."

b. "The numbers in my BINGO add up to 138."

c. "My BINGO is a vertical line."

d. "My BINGO uses the FREE space."

B	I	N	G	O
2	22	32	50	61
14	30	35	47	64
7	25	Free	52	72
5	19	37	59	70
11	27	45	48	75

One statement was false; the rest were true. Where was Jane's BINGO?

28. Each series below follows its own logical rules. Can you determine the next in each series?

a.

    6 (10) (20) 10 (14) (7) (?) (18)

b. 8 13 14 19 20 25 26 ? ? 37 38

c. 12 16 8 12 6 10 5 ?

d. 27 54 18 36 12 24 8 ?

e. 3 6 12 15 18 36 72 75 150 ?

29. Can you find the number that's described by the poem?

> *The number you seek is more than 50*
>
> *But stay under 100 to be more thrifty*
>
> *A powerful number these digits are called*
>
> *A product of square and cube, so scrawled*
>
> *The square is from a simple three*
>
> *You're partly given for free*
>
> *The number to cube is even and close*
>
> *A prime number too, so now it's exposed*
>
> *So take square and cube worked out*
>
> *Make the product and don't you pout!*
>
> *If you've listened to what I've said*
>
> *You know the answer to what you've read.*

30. The average of three numbers is 40. All three are whole positive numbers and are different from each other.

If the lowest is 19, what could be the highest possible number of the remaining two numbers?

31. What comes next in this sequence:

A1   B2   D4   G7   K11   P16   ??

Just for Fun: Frame Game

32. Find the hidden word or phrase.

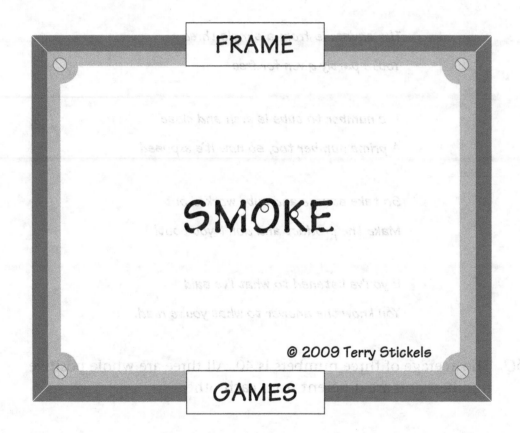

FRAME

SMOKE

© 2009 Terry Stickels

GAMES

33. Casey is making a numbered wheel for a game he and his friends are going to play. His friend Maggie comes by and says, "I see the number 10 is opposite 25 and the number 1 is opposite 16. Did you know there is a simple way to tell how many objects are on an evenly spaced circle if there is an even number?"

HINT
Try examples with smaller numbers.

Casey said he thought he knew. Can you help him out? How many numbers are in the circle Casey is making?

34. One of these numbers doesn't belong with the rest. Which one?

252   333   81

9   240

27

35. What are the missing numbers?

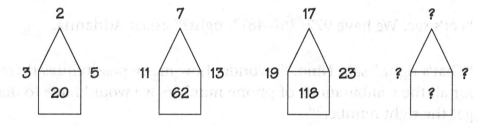

36. A survey showed that of 100 high school students, 50 of them took biology, 20 took chemistry, and 12 took both. How many of the 100 students took neither biology nor chemistry?

37. Richey bought a pack of gum and received 80 cents back in change. The clerk gave him all nickels and dimes. Richey's friend Ben said, "There are seven possible ways you could receive 80 cents in dimes, nickels or a combination of both."

Ben's sister said, "You're wrong, but the correct answer is an odd number."

How many ways are there, and what are they?

38. Nino and Adrianna were trying to remember their cousin's phone number in Dallas. "I know it is 972 area code," said Nino. Adrianna continued ". . . and I know the next two numbers are 36, but I don't know the third number after 6."

"I also know that the last four numbers begin with 48. But I don't know the last two numbers of that group," Nino said, scratching his head.

"Let's see. We have 972–36?–48??, right?" asked Adrianna.

"That's right," said Nino. "I wonder how many possibilities there are for all the combinations of phone numbers we would have to dial to get the right number?"

Adrianna said, "I think there's an easy way to find out if we just think about it logically."

Can you come up with the number of possibilities the phone number might be? Remember—you have to consider "00" as a number, too.

39. Which number comes next in the following sequence?

24, 68, 101, 214, 161, 820, _?_

    a. 1,000

    b. 1,002

    c. 222

    d. 14

40. If two typists can type two pages in 4 minutes, how many typists will it take to type 10 pages in 20 minutes?

    a. 2

    b. 4

    c. 8

    d. 16

    e. 20

41. In a bass fishing tournament, 200 bass were caught in 5 days. The total fish caught on each day was 8 more than the day before. How many fish were caught on the first day?

42. Below is a target that indicates the scores the arrows can make.

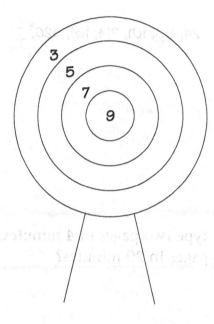

Six of the scores below are possible total scores if you shoot four arrows. Two scores are not. Which two scores are impossible?

Scores:	24	22
	34	32
	30	21
	33	26

43. Two numbers with no zeros in their make-up can be multiplied to create 10,000. They are 625 and 16. Is it possible to have two numbers multiplied together with no zeros that equal 100,000? 1,000,000?

HINT
Use smaller examples like 100 and 1,000. It's also OK to use a calculator.

Just for Fun: Frame Game

44. Find the hidden word or phrase.

FRAME

P*a*ce

© 2009 Terry Stickels

GAMES

45. Here are three puzzles that are easier to solve than they first appear, if you know the right path to their solutions.

a. What is the remainder when 10^{93} is divided by 9?

b. What is the remainder when 4^{69} is divided by 10?

c. Two of the numbers below are divisible by 9 with no remainder. There is a quick way to find out which two if you know some simple rules of divisibility. Which two?

111,111   222,222   333,333

444,444   555,555   666,666

777,777   888,888

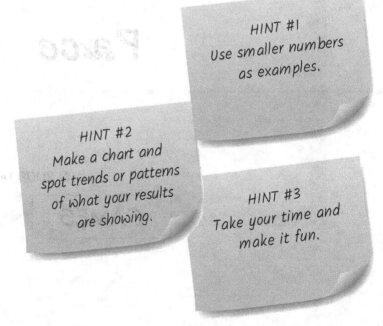

HINT #1
Use smaller numbers as examples.

HINT #2
Make a chart and spot trends or patterns of what your results are showing.

HINT #3
Take your time and make it fun.

46. What is the next number in the sequence below?

142,857

285,714

428,571

571,428

714,285

???,???

a. 785,241

b. 587,421

c. 875,421

d. 857,142

47. The average of four positive integers less than 10 is 8.

a. No number can be less than ______________.

b. What different combinations of the four integers will give you an average of 8?

HINT
How do you know for certain that four numbers will average 8?

48. How many 40's must be added together to get a sum equal to 40^4?

    a. 1,000

    b. 1,000,000

    c. 64,000

    d. 64,000,000

HINT
Use examples with smaller numbers.

49. Here is an example of pairing numbers.

A	1	2	3	4	5	6	...	13	20	60
	↓	↓	↓	↓	↓	↓		↓	↓	↓
B	3	6	9	12	15	18	...	39	60	180

In this example, Row B is three times each respective number in Row A. Now find the missing numbers in the number pairing puzzles below.

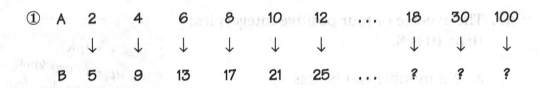

① A	2	4	6	8	10	12	...	18	30	100
	↓	↓	↓	↓	↓	↓		↓	↓	↓
B	5	9	13	17	21	25	...	?	?	?

② A	1	2	3	4	5	6	...	10	15	30
	↓	↓	↓	↓	↓	↓		↓	↓	↓
B	1	4	9	16	25	36	...	?	?	?

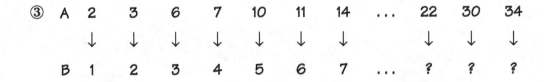

③ A	2	3	6	7	10	11	14	...	22	30	34
	↓	↓	↓	↓	↓	↓	↓		↓	↓	↓
B	1	2	3	4	5	6	7	...	?	?	?

50. The numbers 10, 12, 14, 16, and 18 comprise a set of five consecutive even numbers.

Suppose the sum of five consecutive numbers is 640. What are the five numbers?

HINT #1
What might be a logical beginning point to find the numbers?

HINT #2
Because there is an odd number of consecutive numbers, what could the middle number be?

51. Here's a follow-up to the last puzzle. What if I ask you to list the six consecutive numbers that total 630? It's an even number of consecutive numbers, so you'll have to make a slight adjustment in your logic, but it's not difficult. Why do you think I chose 630 instead of 640?

HINT
Whole numbers only!

52. The mobile below is perfectly balanced. The two upside-down triangles each weighs 20 pounds. If the hourglass weighs 14 pounds and the star weighs 4 pounds, what is the weight of each box and the circle?

HINT
The side with the hourglass has to also balance the side with the circle and star.

Hourglass = 14 pounds
Star = 4 pounds

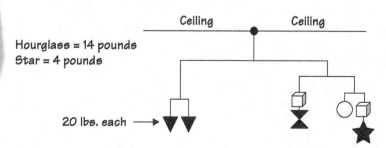

20 lbs. each ⟶

53. A Fazooto and a Pazooto together cost 50 cents. A Pazooto and a Razooto together cost 60 cents. A Razooto and a Fazooto together cost 70 cents. How much does each Pazooto cost?

54. Each one of the four circles below (A, B, C, D) has a value somewhere between 1 and 9 (including 1 and 9). No two circles have the same value. The number 21 is the sum of the values of circles A, B, and C. The number 14 is the sum of the values of C and D.

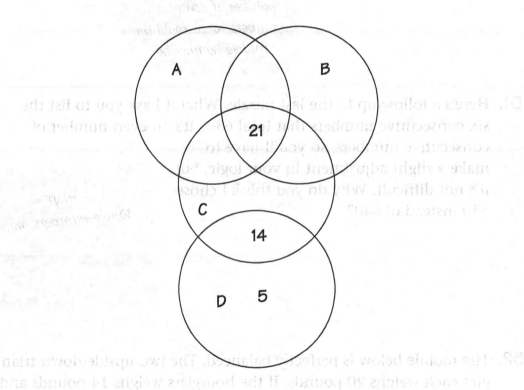

What are the values of circles A and B?

Just for Fun: Frame Game

55. Find the hidden word or phrase.

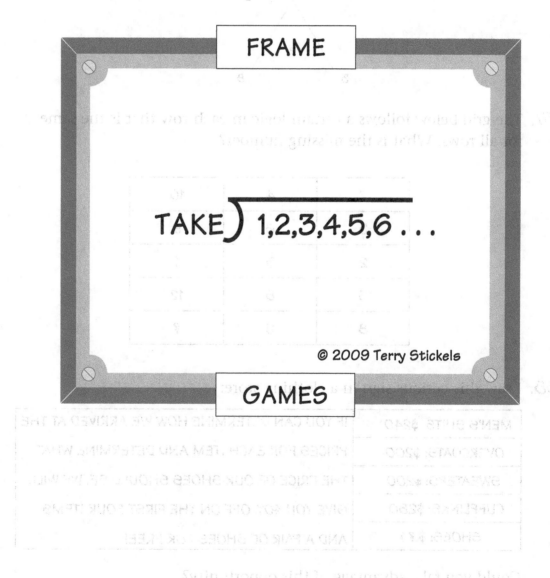

FRAME

$$\text{TAKE}\overline{)\ 1,2,3,4,5,6\ \ldots}$$

© 2009 Terry Stickels

GAMES

56. What is the missing number?

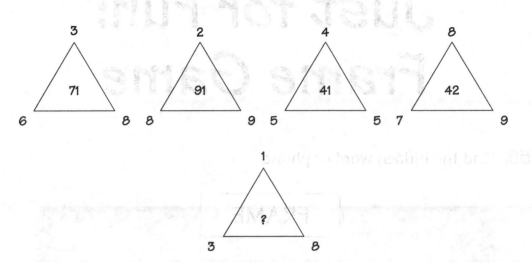

57. The grid below follows a certain logic in each row that is the same for all rows. What is the missing number?

7	4	10
3	6	9
2	5	1
6	8	12
8	9	?

58. I saw this curious sign in a clothing store:

MEN'S SUITS: $240	IF YOU CAN DETERMINE HOW WE ARRIVED AT THE
OVERCOATS: $200	PRICES FOR EACH ITEM AND DETERMINE WHAT
SWEATERS: $200	THE PRICE OF OUR SHOES SHOULD BE, WE WILL
CUFFLINKS: $280	GIVE YOU 40% OFF ON THE FIRST FOUR ITEMS
SHOES: $??	AND A PAIR OF SHOES FOR FREE!

Could you take advantage of this opportunity?

59. What number multiplied by itself is the product of 36 × 196?

60. There is an old puzzle about a student who takes a 10-question test. She receives 5 points for each correct answer and has 2 points taken away for each incorrect answer. She answers all 10 problems and receives a score of 29. The puzzle then asks how many questions she answered correctly. The problem is that these puzzles give the answer but no explanation. Will you give it a try?

HINT
Make a chart of all possible scores.

61. The numbers 1–16 are to be placed in a magic square so that each of the rows, columns, and diagonals has the same sum. What is x?

		3	16
___	15	_x_	5
14	___	8	11
7	12	13	___

HINT #1
Look at other rows, columns, and diagonals to see what numbers are left to use.

HINT #2
Once you find x, you'll find the others easily.

62. Suppose all the counting numbers are arranged in columns as shown below:

A	B	C	D	E	F	G	H
1	2	3	4	5	6	7	8
9	10	11	12	13	14	15	16
17	18	19	20	21	22	23	24
25	26	--	--	--	--	--	--

Under what column will 1,000 appear?

63. The numbers below are grouped using a certain logic that is the same for all six rows.

$$309 - 64 - 42$$
$$741 - 82 - 15$$
$$282 - 37 - 06$$
$$552 - 19 - 33$$
$$183 - 55 - 24$$
$$624 - 28 - ?$$

Which of the numbers below could be the missing number?

a. 53

b. 34

c. 51

d. 61

Copyright © 2009 by John Wiley & Sons, Inc.

64. A pet lizard doubled in length each year until it reached its maximum length over the course of 12 years. How many years did it take for the lizard to reach half its maximum length?

65. My young brother's age today is 2 times what it will be 2 years from now minus 2 times what his age was 2 years ago. Is he less or more than 10 years old? Can you find his exact age now?

Just for Fun: Frame Game

66. Find the hidden word or phrase.

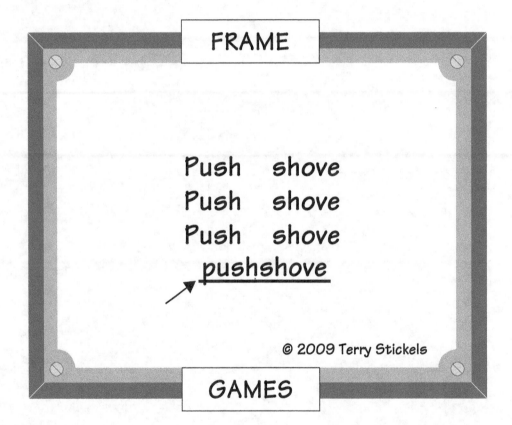

© 2009 Terry Stickels

67. Quick now, if you write down all the numbers from 400 to 500, how many times will you write the digit 4?

68. Tom and Trey were playing a ball toss game. The two values in the ring were 5 points and 8 points. Trey said, "I wonder what's the highest score that *can't* be made using any combination of 5 and 8."

Tom replied, "I don't think it could be very high because I know you can make every number from 40 and higher using different combinations of 5's and 8's."

What is the highest number that *can't* be formed using only 5's and 8's?

69. What is the sum of the numbers in the sixth row, and what are the middle two numbers in that row?

			1				Row 1
		1		1			Row 2
	1		2		1		Row 3
1		3		3		1	Row 4
1	4		6		4	1	Row 5
?	?	?		?	?	?	Row 6
1	6	15	20	15	6	1	Row 7

70. If you saw an arithmetic sequence with the consecutive numbers listed below . . .

$$. . . 20, 24, 28, 32, 36, 40, 44, 48 . . .$$

. . . which of the following terms would be a possible number in the sequence?

a. 393

b. 838

c. 262

d. 756

71. The numbers in columns X and Y have been manipulated in a logical way to create the numbers in the Results column. The results are determined horizontally.

X	Y	Results
4	3	5
8	6	34
2	9	7
6	1	−1
10	5	35
9	7	?

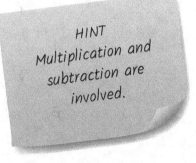

HINT
Multiplication and subtraction are involved.

What is the result in the last row?

72. Brad had three new tires for his bike and decided to rotate them for equal use. He rotated them every 3,000 miles. How many miles did each tire have after 24,000 miles?

73. A certain logic has been used to determine the middle number in each diamond. What is the missing number?

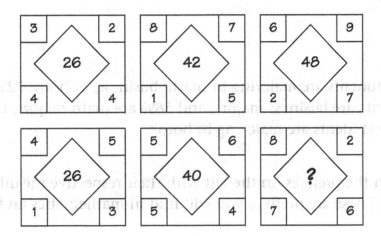

74. How many numbers are in this sequence if all the missing numbers indicated by "..." are included?

0, 3, 6, 9, 12, 15, 18, 21, 24, . . . 900?

HINT
Make a chart of the position of the numbers.

75. Using the numbers 0, −1, −4, −5, −7, −8, and −9, can you fill in the remaining seven boxes so each row, column, and diagonal has the same total?

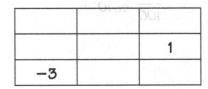

Rational Numbers

76. 100 students are majoring in math, business, or both. 72% of the students are business majors, and 58% are math majors. How many students are majoring in both?

77. Match the prefixes on the left with their respective meanings on the right. These are prefixes you will find in mathematics and sciences.

a. deci-   1,000,000 or 10^6

b. centi- 10^9

c. giga- 10^{-12}

d. hecto- $\dfrac{1}{10}$ or 10^{-1}

e. kilo- $\dfrac{1}{1,000}$ or 10^{-3}

f. mega-   100 or 10^2

g. milli-   1,000 or 10^3

h. pico- $\dfrac{1}{100}$ or 10^{-2}

Just for Fun: Frame Game

78. Find the hidden word or phrase.

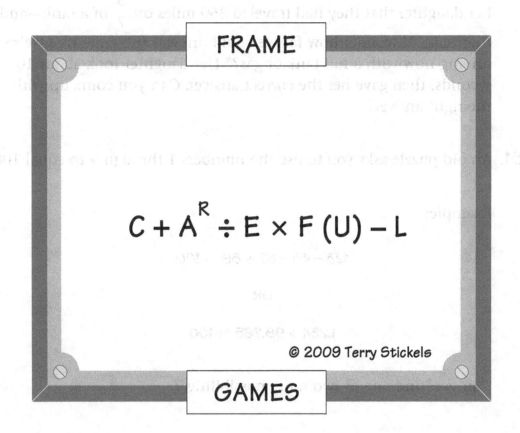

FRAME

$$C + A^R \div E \times F (U) - L$$

© 2009 Terry Stickels

GAMES

79. Can you create the number 2 by using four 4's? As an example, you can make 17 out of four 4's like this: $(4 \times 4) + \dfrac{4}{4} = 16 + 1 = 17$.

You can use any math symbols you choose for this puzzle.

80. A mother fills up her gas tank and realizes she had used only $\dfrac{5}{7}$ of the gas before she pulled into the service station. She remarked to her daughter that they had traveled 360 miles on $\dfrac{5}{7}$ of a tank—and then said, "I wonder how far we could drive at the same rate we're driving now with a full tank of gas?" Her daughter took about 10 seconds, then gave her the correct answer. Can you come up with the right answer?

81. An old puzzle asks you to use the numbers 1 through 9 to equal 100.

Example:

$$123 - 45 - 67 + 89 = 100$$

OR

$$1.234 + 98.765 = 100$$

Can you find one or two more possibilities?

82. It is possible to create all the numbers from 1 to 100 using just four 4's—and, of course, a combination of math symbols and operations. Try your luck at creating 42.

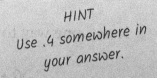

HINT
Use .4 somewhere in your answer.

83. What is the relationship between:

$$A \quad \frac{1}{10\sqrt{10}} \quad \text{and} \quad B \quad \frac{\sqrt{10}}{\sqrt{10^4}}$$

a. A is larger.

b. B is larger.

c. They are equal.

84. What is the value of $\dfrac{10^{11}+10^{10}}{10^{10}}$?

a. 10

b. 10^{21}

c. 11

d. 2^5

85. Puzzles with different number bases are fun because there are different options for new numbers. Here's an example:

What is the number 35 in $Base_9$?

Here's one way to approach this puzzle:

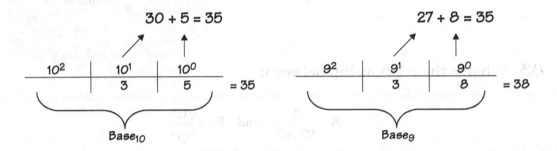

Try these:

a. Convert 113 in $Base_{10}$ to $Base_7$.

b. Convert 262 in $Base_{10}$ to $Base_5$.

c. Convert 534 in $Base_6$ to $Base_4$.

86. See how quickly you can come up with the answers to these analogies.

Example:

Black : White : : Night : _?_   Answer: Day.

a. $270° : \dfrac{3}{4} : : 300° :$ _?_

b. Hexagon : Triangle : : _?_ : Square

c. 3 : 9 : 81 : : 5 : 25 : _?_

87. If 1 dollar in Konklo currency is equal to 4.28466 Ponplos, then 1 Ponplo equals ___ Konklos.

Just for Fun: Frame Game

88. Find the hidden word or phrase.

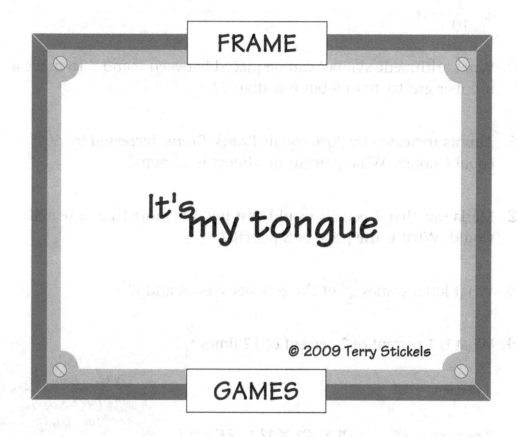

FRAME

It's my tongue

© 2009 Terry Stickels

GAMES

89. Which one of the following has the largest value?

    a. .0001

    b. $\dfrac{1}{1000}$

    c. $(-.100)^2$

    d. $\dfrac{1}{10^4}$

    e. $\dfrac{1}{10} \div .1$

90. What arithmetic symbol can be placed between 6 and 7 to make a number greater than 6 but less than 7?

91. Almors increased by 20% equals Brons. Brons decreased by 50% equal Choops. What percent of Almors is Choops?

92. Maria saw that 7 pencils would cost her 30¢ more than 5 pencils would. What is the price of 8 pencils?

93. What letter comes $\dfrac{2}{5}$ of the way between A and J?

94. What is 1 percent of 2 percent of 12 times $\dfrac{2}{3}$?

1 percent = .01

2 percent = .02 $.01 \times .02 \times 12 \times .666 = \underline{\textbf{?}}$

$\dfrac{2}{3} = .666\ldots$

> HINT
> It might help to set this puzzle up by listing what you know.

95. Using the numbers 0, 1, 2, 3, 4, and 9 once and only once, make two numbers with decimals whose difference is between 52 and 53. There may be more than one answer.

96. Without using a calculator, decide which of the following has the largest quotient.

a. $\dfrac{27}{.04}$

b. $\dfrac{.27}{.40}$

c. $\dfrac{2.7}{.04}$

d. $\dfrac{27}{40.0}$

e. $\dfrac{27}{.004}$

97. 300 percent more than 60 is _____.

98. A particle floats with $\dfrac{5}{7}$ of its weight above the surface. What is the ratio of the particle's submerged weight to its exposed weight?

Just for Fun: Frame Game

99. Find the hidden word or phrase.

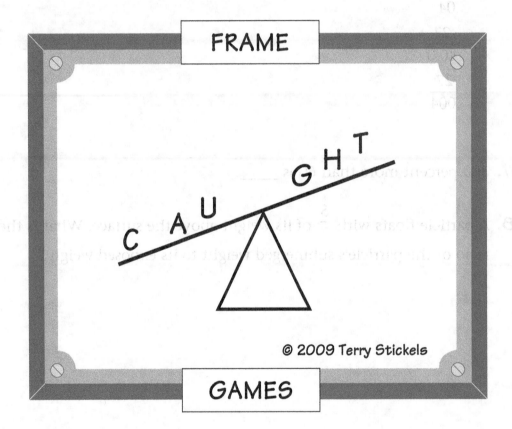

FRAME

C A U GHT

© 2009 Terry Stickels

GAMES

100. In Mike's last baseball game, he threw 33 strikes and 11 balls. What percentage of Mike's pitches were strikes? Of the 44 total pitches, how many strikes would Mike have had to throw to reach 85%?

101. Millie's sister told her she could win a chocolate malt if she could solve the following puzzle in 30 or fewer seconds. Can you?

$$\text{What is } \frac{1}{3} \div \frac{13}{11} \div \frac{4}{7} ?$$

    a. Close to $\dfrac{1}{2}$

    b. Close to 17

    c. Close to 10

    d. Close to $\dfrac{1}{11}$

102. Using the numbers 3, 4, 5, and 6 once and only once, what two fractions can you write where .8 will lie between them?

103. Let's see how well you know the rules of exponents.

    a. $? = \sqrt{\sqrt{\sqrt{256}}}$

    b. $\left(\left(X^{\frac{1}{2}} \right)^{\frac{1}{2}} \right)^{\frac{1}{2}} = ?$

104. What is the next number in the sequence below?

$$\frac{1}{3} \quad \frac{1}{2} \quad \frac{1}{5} \quad \frac{1}{4} \quad \frac{1}{7} \quad \frac{1}{8} \quad \frac{1}{9} \quad \frac{1}{16} \quad \frac{1}{11} \quad \frac{1}{32} \quad \frac{?}{\underline{}}$$

105. Sam works in the stock room at Widget Factory. This morning he got an order for gidgets and gadgets, but he can't fill the order because it has a problem. Due to some networking problems with the company's servers, the order is missing some key information. Somehow a bunch of the numbers got turned into Xs, and he can't quite figure out what the order is for. Can you help him figure out what the missing numbers are?

Ship To:	Mary Ellsbeth 16 Forest Drive Townsville, Alaska 02671		DATE: January 3, 2010
Quantity	Description	Unit Price	Total Price
2	Brite-Lite Gadget	$ XX.XX	$ 32.50
X	Useful Gidget	$ 7.30	$ X.XX
2	Gadget Attack! Game	$ 12.95	$ XX.XX
X	Gidget Packs	$ 4.85	$ 14.55
		Sub-Total	$ XX.XX
		Shipping:	$ 8.00
		Total Order:	$ 88.25
		Total Due:	Paid in Full

106. 7 numbers have an average of 6. 6 numbers have an average of 7. 3 numbers have an average of 42. What is the average of all 16 numbers?

a. $12\frac{1}{2}$

b. $40\frac{1}{3}$

c. $13\frac{1}{8}$

d. 18

107. What is 30% of $\frac{25}{17}$ divided by $\frac{3}{4}$ times $\frac{1}{2}$?

108.

A third of a third
Won't make you a nerd.

A half of a half
Might get you a laugh.

A fifth of a fifth
Will dispel the myth.

That math can't be fun.

So no more rhymes
No more tease

Just add these all up
If you please.

Tell me the sum
If you'd be so kind.

It will ease my tension
And soothe my mind.

109. Suppose today is Tuesday. What day of the week will it be 200 days from now?

HINT
Think of a diagram or chart to make the puzzle easier.

Just for Fun: Frame Game

110. Find the hidden word or phrase.

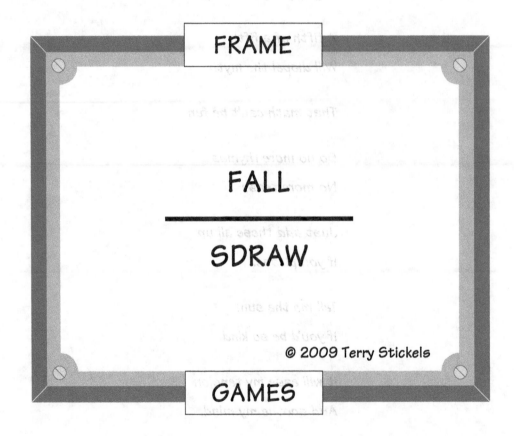

FRAME

FALL

―――

SDRAW

© 2009 Terry Stickels

GAMES

111. If $\frac{1}{2}$ of a zeeko is 60% of a teeko, then a zeeko divided by a teeko is:

a. 30

b. 1

c. $\frac{5}{6}$

d. 1.2

112. One of the following fractions is less than $\frac{1}{9}$. Which one?

a. $\frac{31}{200}$

b. $\frac{19}{211}$

c. $\frac{17}{150}$

d. $\frac{113}{983}$

113. If the Celtics win 40% of their games in the first third of their season, what percent of the remaining games would they have to win so they can finish the season with an even record (50% wins)?

114. A increased by 30% = B

B decreased by 40% = C

C increased by 20% = D

What percent of B is D?

115. In each pair shown below in items a, b, and c, which of the two is larger, or are they the same? (No calculators please!)

a. 5^{10} or 10^5

b. $\dfrac{\dfrac{1}{\sqrt{.1}}}{\sqrt{.01}}$ or $\dfrac{\dfrac{1}{\sqrt{.01}}}{\sqrt{.1}}$

c. 3^{12} or 9^6

116. Can you express this fraction in lowest terms?

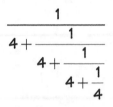

$$\cfrac{1}{4+\cfrac{1}{4+\cfrac{1}{4+\cfrac{1}{4}}}}$$

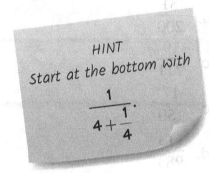

HINT
Start at the bottom with $\dfrac{1}{4+\dfrac{1}{4}}$.

Here are some choices:

a. $\dfrac{1}{5}$

b. $\dfrac{24}{101}$

c. $\dfrac{27}{102}$

d. $\dfrac{1}{3}$

117. Three friends form a company and agree to share the profits based on the proportion of the amount invested. For example, if two people formed a partnership and one invested $2,000 and the other invested $1,000, the first partner would be paid double the profits of the second partner.

The three partners invested $3,000, $4,000, and $5,000. In the first year, the company made a profit of $60,000. How much did each partner receive, based on the original investment?

118. One of the following fractions does not fit the pattern set by the others. Which is the odd one out, and what should the correct fraction be?

$$\frac{1}{2} \quad \frac{1}{3} \quad \frac{1}{4} \quad \frac{1}{6} \quad \frac{1}{9} \quad \frac{1}{12} \quad \frac{1}{19} \quad \frac{1}{36}$$

HINT
Factors.

119. A group of math students were making copies of some documents when the copy machine broke. They fixed the machine but noticed that the 50% and the 10% buttons for reproducing copies were permanently disabled. They were left with these choices for document size:

5%   125%   200%

They had to make a 50% document by using combinations of the remaining three percentage buttons. Can you help them make a copy at 50% in five steps, using at least three different buttons at least once?

120. What is the answer to the following puzzle . . . in fractional form?

$$\frac{1}{4\frac{3}{7}} + \frac{1}{3\frac{11}{13}} + \frac{1}{5\frac{1}{9}} = \frac{?}{\underline{\hspace{1cm}}}$$

Just for Fun: Frame Game

121. Find the hidden word or phrase.

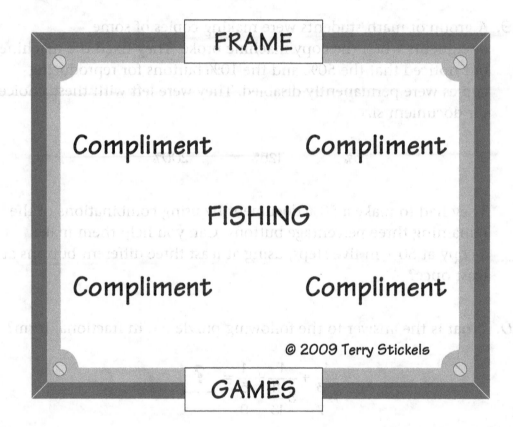

FRAME

Compliment   Compliment

FISHING

Compliment   Compliment

© 2009 Terry Stickels

GAMES

122. Simi was helping his mother prepare drinks for dinner. He was measuring lemonade and placing it into two different-sized containers.

He filled the first container $\frac{1}{3}$ full of lemonade and another container twice the size $\frac{1}{4}$ full of lemonade. He then filled each container with iced tea and emptied the contents of both into a large pitcher. Lemonade and iced tea make a great-tasting summertime drink called an "Arnold Palmer."

Simi's mom then said, "You're pretty good in math, so tell me what part of the new mixture is lemonade and what part is iced tea?" Can you help Simi find the answer?

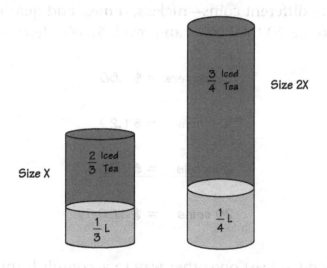

123. Penny conducted a recent poll in her hometown. She found that 80% of all the people in her city really liked chocolate. She also found that 12% of the town's residents were under the age of 15.

Her brother asked her, "If you were to pick anyone at random, what are the chances that person didn't like chocolate and was older than 15?"

124. $\dfrac{50\% \times 200\% \times 100\%}{25\%} = \underline{\textbf{?}}$

125. How many numbers between 1 and 100 can divide into 109 and have a remainder of 5? List those numbers. Is there an easy way to find them?

126. Using three different coins—nickels, dimes, and quarters—it is possible to use 20 total coins and reach $2.00. Here's one way:

$$2 \text{ quarters} = \$ \ .50$$

$$12 \text{ dimes} = \$1.20$$

$$\underline{6 \text{ nickels} = \$ \ .30}$$

$$20 \text{ coins} = \$2.00$$

Can you find at least one other way to accomplish this?

127. Mimi and Suzanne wanted to buy an ice cream cone, but they realized that they were both short of having enough money. And when they put their money together, they still didn't have enough money for one cone. Mimi was 53 cents short, and Suzanne was 48 cents short. What is the least the ice cream cone could cost?

Just for Fun: Frame Game

128. Find the hidden word or phrase.

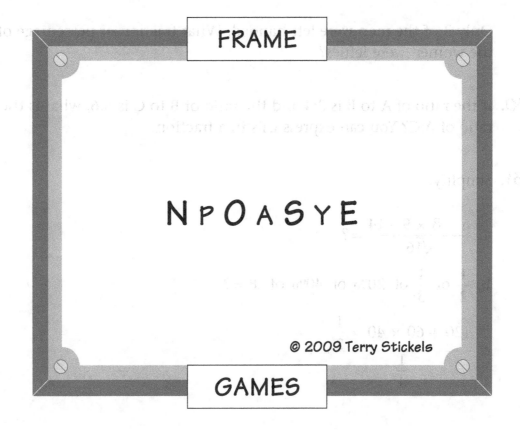

FRAME

N p O A S Y E

© 2009 Terry Stickels

GAMES

129. There were 100 men and women at a baseball card show:

59 were men.

72 were football and baseball card collectors.

81 lived in the city.

89 were right-handed.

Only 3 of the men were left-handed. What fraction or percentage of the women were lefties?

130. If the ratio of A to B is 3:4 and the ratio of B to C is 5:6, what is the ratio of A:C? You can express this in a fraction.

131. Simplify:

a. $\dfrac{6 + 8 \times 9 - 14}{\sqrt{16}} = ?$

b. $\dfrac{1}{2}$ of $\dfrac{1}{3}$ of 20% of 40% of .8 = ?

c. $\dfrac{120 + 60 \times 40 \times \dfrac{1}{2}}{\dfrac{1}{4}} = ?$

132. A swimming pool is 100 ft. by 50 ft. and has an average depth of 6 ft. By the end of an average summer day, the water level drops an average of 1 ft.

How many cubic feet of water are lost each day?

133. Four-fifths of a math class is made up of female students. What is the ratio of male students to female students?

134. What is the percentage of all integers that contain at least one 4?

135. A drain can empty $\frac{5}{8}$ of a sink in one minute and the faucet can fill $\frac{3}{4}$ of the sink in one minute. If the faucet is on and the drain is open, how long will it be before the sink is full?

136. A young boy was asked to measure off and mark 6 ft. exactly on a sidewalk. All he had was a piece of rope and a shed that measured 5 ft. by 7 ft. Can you think of a way he might be able to accomplish this?

137. Jane bought a $40 pair of jeans and received a 20% discount, and then bought a $30 shirt and received a 40% discount when she checked out. What is the single percent discount for the total purchase?

Just for Fun: Frame Game

138. Find the hidden word or phrase.

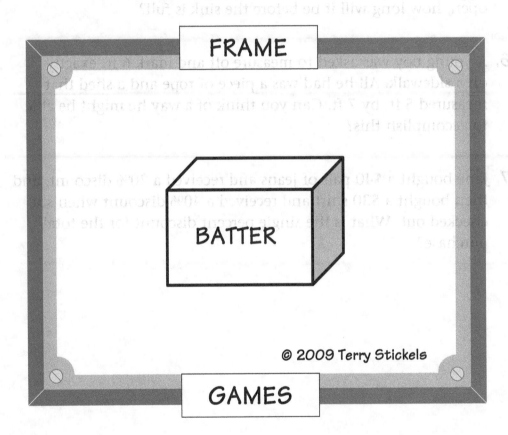

FRAME

BATTER

© 2009 Terry Stickels

GAMES

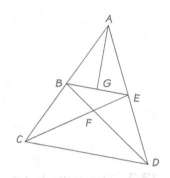

Part II

GEOMETRY
and
MEASUREMENT

Geometry

139. The cube below can be unfolded into one of the four patterns A–D. Which is the correct choice?

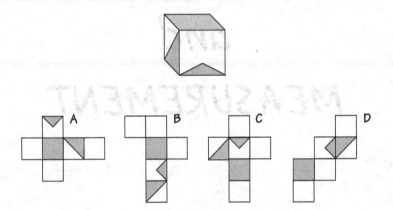

140. Below are five versions of an unfolded cube. One of the choices is not possible. Which one?

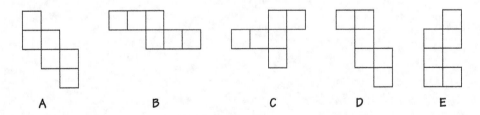

141. Two of the four boxes below can be folded into the cube at the right. Which ones?

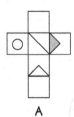 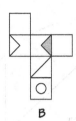 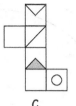

A   B   C   D

142. Below are drawings of different hoses to fill swimming pools. Illustration A is a combination of two hoses, each having a 16-inch radius. The hose in illustration B has a 30-inch radius. If the force of the flow of water coming out of the hoses is the same in both A and B, which will fill a swimming pool faster?

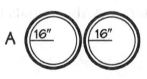   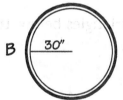

A   16″   16″   B   30″

143. If you were to draw the following figures, one would be noticeably different from the rest, based on a simple, straightforward design characteristic. Which figure would be the odd one out?

a. Parallelogram

b. Trapezoid

c. Hexagon

d. Quadrilateral

e. Rectangle

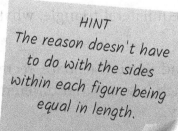

HINT
The reason doesn't have to do with the sides within each figure being equal in length.

144. Rectangle EFHG is inscribed in the circle below. Line AE represents 5 feet. Line EG represents 7 feet. What is the length of Line EH?

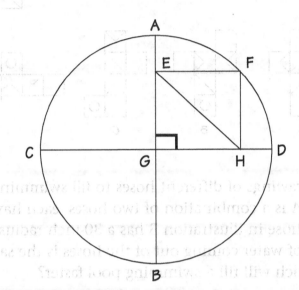

145. For the triangles below, the perimeter of △ABC equals the perimeter of △RST.

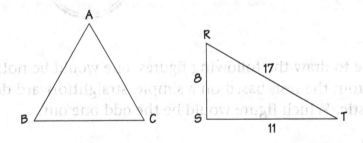

If △ABC is an equilateral triangle, what is the length of AB?

146. Of the four quadrilaterals (square, rectangle, rhombus, and trapezoid), which is somewhat different from the others because of its design? Why?

147. Two of the four illustrations below will fold into the prism shown in the middle. Which two?

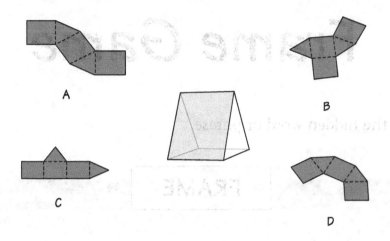

148. Billings Street runs due north from Rucker 4 blocks, where it meets Genco Blvd., an east–west street. If you go due east on Genco Blvd. for 13 blocks and turn right, you cross Rucker, then Lawford, and come to Baylor, which goes west only. If you drive 6 blocks west on Baylor and turn right, you'll be on Saymore St.

The street you turned onto off of Genco Blvd. is:

a. 90° to Saymore St.

b. parallel to Saymore St.

c. runs across Billings St.

d. Can't tell from the information

Just for Fun: Frame Game

149. Find the hidden word or phrase.

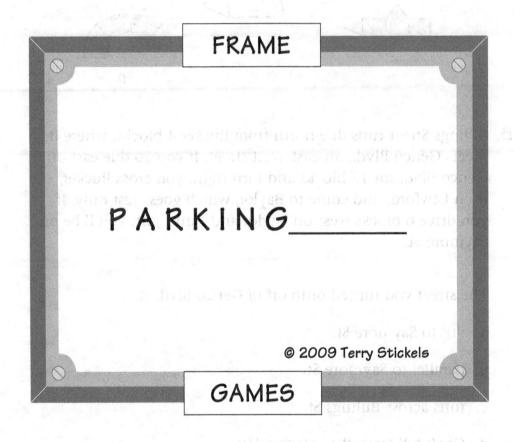

FRAME

PARKING_____

© 2009 Terry Stickels

GAMES

150. How many four-sided figures are in the drawing below?

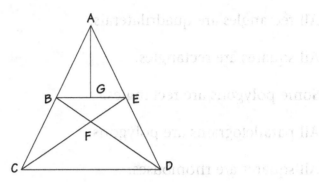

151. What is the angle of the question mark below?

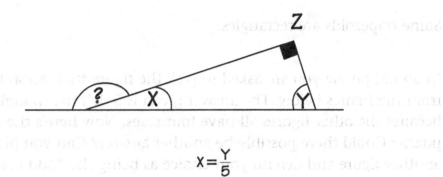

$$X = \frac{Y}{5}$$

152. Grids A and B are identical. Which line segment is longer—A or B?

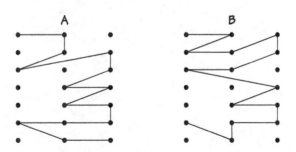

153. Which of the following statements are true?

All rectangles are quadrilaterals.

All squares are rectangles.

Some polygons are rectangles.

All parallelograms are polygons.

All squares are rhombuses.

Some rectangles are rhombuses.

All trapezoids are quadrilaterals.

Some trapezoids are rectangles.

154. In an old puzzle you are asked to pick the figure that doesn't belong from the figures below. The answer given is 3: It's the triangle, because the other figures all have four sides. Now here's the new puzzle: Could there possibly be another answer? Can you pick another figure and defend your choice as being the "odd man out"?

155. Marian's class was going to have a pizza party and she was responsible for ordering the pizza. Louie's Pizza makes square pizzas and sells an 18″ × 18″ pizza for $15.00 as a special. Dominick's Pizza House has the normal circular pizza, which is 20″ in diameter. It's on special for $16.00. Which pizza is the better deal?

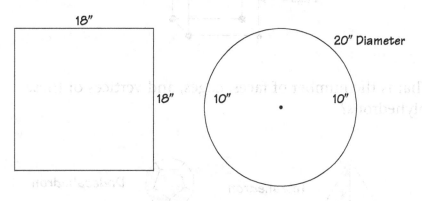

156. The lengths of two sides of an isosceles triangle are 7 inches and 20 inches. What is the perimeter of the triangle?

157. If P is a point on a line called XY (X is the left-most point and Y is the right-most point), which of these must be true?

a. XP = PY

b. XP = XY – PY

c. XY = XP – PY

d. PY = XP + PY

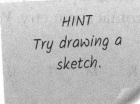

HINT
Try drawing a sketch.

158. A cube has 6 faces, 12 edges, and 8 vertices.

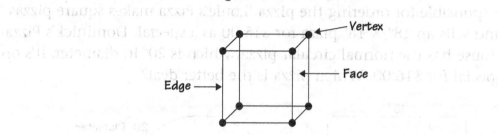

What is the number of faces, edges, and vertices of these polyhedrons?

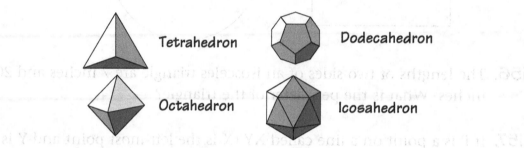

159. All except one of these capital letters has vertical symmetry. The odd one out has horizontal symmetry. Which one?

A   K   M   T   U   V   W   Y

160. If you write a capital "G" backward and turn it upside down in a mirror, it will appear like which of the following:

161. In 2 hours and 40 minutes the minute hand of a clock rotates through an angle of:

a. 90°

b. 96°

c. 960°

d. 720°

Just for Fun:
Frame Game

162. Find the hidden word or phrase.

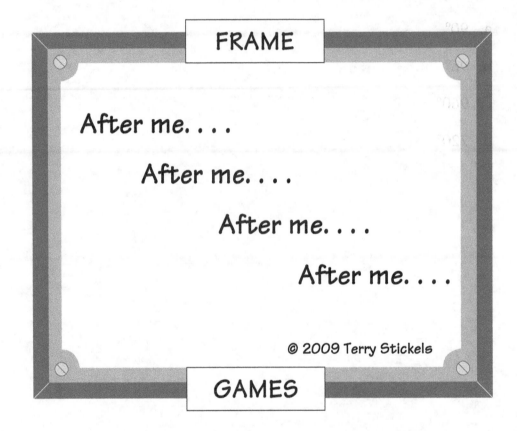

FRAME

After me. . . .

    After me. . . .

        After me. . . .

            After me. . . .

© 2009 Terry Stickels

GAMES

Measurement

163. The times on the clocks below are in a logical sequence of times from 1 to 8. What time should be on the seventh clock?

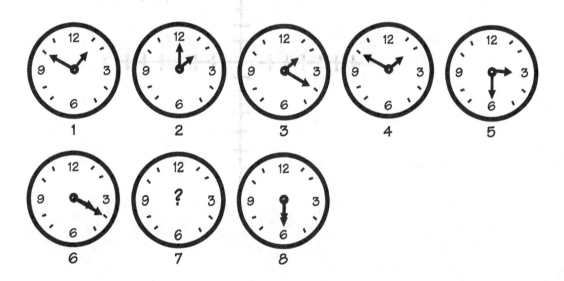

164. If I wanted to lay carpeting in a 40 × 18 foot room, how many square yards of carpet would I need? (3 feet = 1 yard)

165. The point (3, 4) is shown on the graph below. If you draw a straight line from the origin (0, 0) to (3, 4), the length of that line is:

a. less than 4 units

b. less than 3 units

c. greater than 4 units

d. exactly 4 units

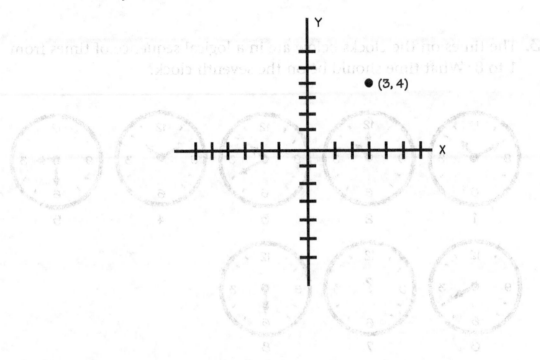

166. The diagram below shows a small pulley pulling around a larger driver pulley.

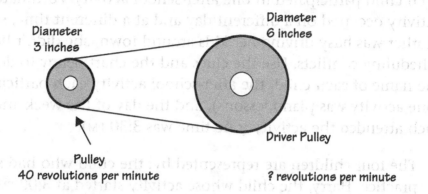

Diameter
3 inches

Diameter
6 inches

Driver Pulley

Pulley
40 revolutions per minute

? revolutions per minute

If the 3-inch pulley makes 40 revolutions per minute, the 6-inch pulley rotates:

a. 10 times

b. 20 times

c. 40 times

d. Can't tell from the picture

167. Below is a plank that has to be balanced by two weights on either side of a fulcrum.

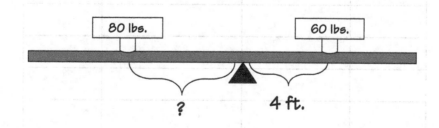

80 lbs.

60 lbs.

?

4 ft.

The 60-lb. weight on the right is 4 feet from the fulcrum. How far from the fulcrum should the 80-lb. weight be placed to balance the plank?

168. Mother just finished setting up her schedule for her children's weekly activities. She has four children: Alicia, Cindy, Harry, and Larry. Each child participated in one after-school activity. Fortunately, each activity occurred on a different day and at a different time, so while Mother was busy driving her kids around town, she didn't have any scheduling conflicts. Use the clues and the chart below to determine the name of each child, the after-school activity each participated in (one activity was piano lessons), and the day of the week and time each attended the activity (one time was 3:30 PM).

a. The four children are represented by: the child who had soccer practice, Harry, the child whose activity started at 3:00 PM, and the child with an activity on Wednesday.

b. Cindy went to her activity on Friday.

c. Larry didn't attend baseball practice, which met on Tuesday.

d. On Thursday, Mother had to take one daughter to her activity at 4:00 PM.

e. The activity that started at 4:30 PM wasn't baseball.

f. One child was taking dance lessons on Wednesday.

Name	Activity	Day of the Week	Time

	Baseball practice	Dance lessons	Piano lessons	Soccer practice	Tuesday	Wednesday	Thursday	Friday	3:00 PM	3:30 PM	4:00 PM	4:30 PM
Alicia												
Cindy												
Harry												
Larry												
3:00 PM												
3:30 PM												
4:00 PM												
4:30 PM												
Tuesday												
Wednesday												
Thursday												
Friday												

169. Mandy thought it would be fun to take an old watch apart. When she put it back together, she accidentally exchanged the hour hand and the minute hand. She had set the watch correctly when it was noon. She left the house and came back at 4:00 PM. What time did the watch read?

170. Max has an old watch he likes, with a large hour hand and 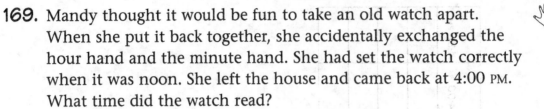 minute hand. He noticed the time as 10:40 when he was walking down the hall to leave school. As he opened the door to leave, he saw the face of his watch upside down and reflected through a plate glass window. The time he saw on his watch reflected back to him was close to:

   a. 9:15

   b. 7:50

   c. 10:20

   d. 3:05

   e. 8:20

171. If the day before yesterday is the 18th, what is two days after tomorrow?

172. John has a clock that runs 12 minutes fast every hour. He and his friend Linda waited until it was noon and then set the clock to 12:00 so it would be correct.

They left for the afternoon and came back, and the clock read 6:00. Linda said, "That's not the real time." John said, "I know. The real time is 5:00. What time will it be when the clock reads 10:48?"

173. A clock with chimes usually chimes on the hour and on the half-hour. They chime the same number of times as the time on the hour and once on the half-hour.

How many total chimes would be heard in a 12-hour period?

Just for Fun: Frame Game

174. Find the hidden word or phrase.

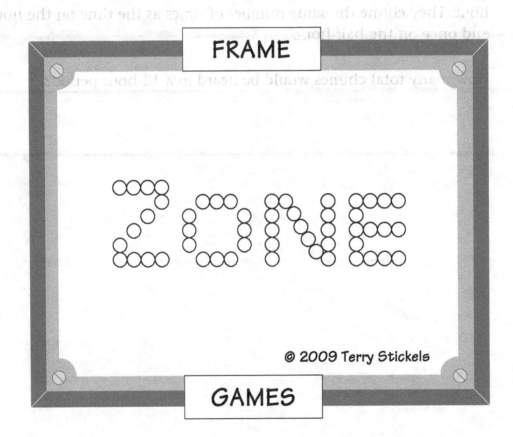

FRAME

ZONE

© 2009 Terry Stickels

GAMES

175. How many days are 100,000 seconds?

176. Ron is trying to figure out how many pieces of chocolate he can cut from an 18″ × 18″ sheet of chocolate if each piece of chocolate is 1.5″ × 2.0″. Can you help him out?

177. Here are some fun math analogies. See how quickly you can come up with the answers.

    a. 100 : Centennial : : 1,000 : ?

    b. Million : Mega : : Millionth : ?

    c. 4 : Square : : ? : Decagon

    d. $\frac{1}{2}$: .5 : : ? : .111111 . . .

    e. 180 degrees : Triangle : : 540° : ?

    f. B : H : : D : ?

      1. G
      2. I
      3. P
      4. X

178. September 21 has 12 hours of daylight and is called the Autumnal equinox. What month has the spring equinox, called the Vernal equinox?

    a. March

    b. April

    c. May

    d. June

179. What's the difference between 1,000 yards square and 1,000 square yards? Can you give an example of what the difference in perimeter might be?

180. Billy and Barbara can mow 1,600 square yards in 81 minutes. How long would it take them to mow 1,600 square feet?

181. In the figure below, ABCD is a square where AB is equal to 10". What is the area of square BDFE?

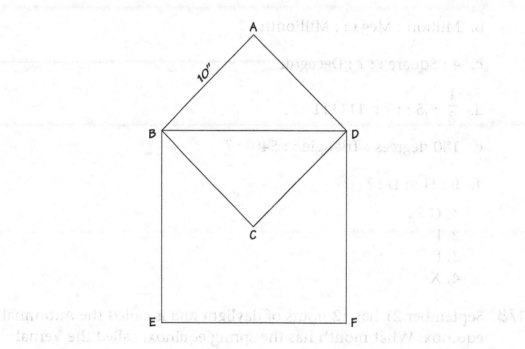

182. Charlie and Rocco are going to a 1:30 PM movie. It takes them 45 minutes to get there from Charlie's home. They are going to stop for a hotdog before the movie, which takes 30 minutes. Charlie's mom asked him to pick up some milk on his way home. That takes another 15 minutes. The movie lasts 2 hours and 15 minutes. If Rocco is at Charlie's home when they leave for the movie and he's going back to Charlie's home after the movie, how long will both boys be gone for the day? What time will the boys arrive home?

183. A clock that doesn't work at all will be correct twice a day, but a clock that loses a minute a day (on a clock that does not show AM or PM) will not be correct again for the following time:

   a. never

   b. 1 year

   c. 720 days

   d. 26 hours

   e. a decade

184. All of the short line segments in this figure are at 90° angles to each other and 1 inch long. What is the total area of the four shaded sections?

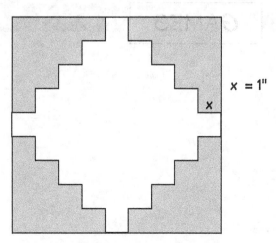

$x = 1''$

Just for Fun: Frame Game

185. Find the hidden word or phrase.

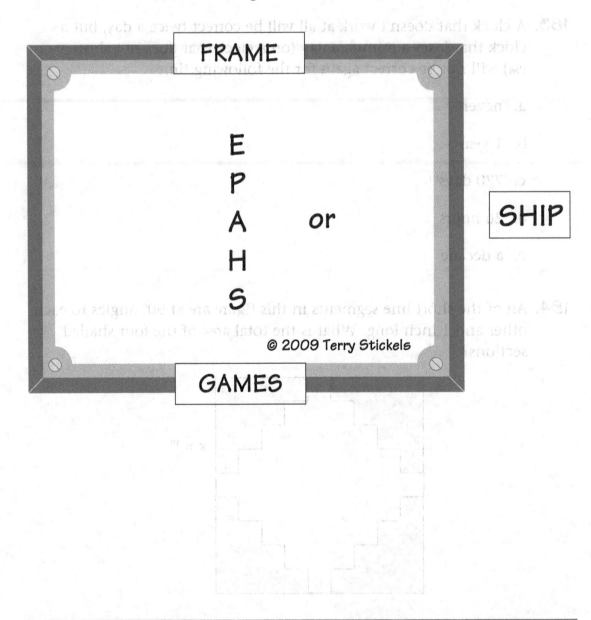

FRAME

E
P
A   or
H
S

SHIP

GAMES

© 2009 Terry Stickels

186. The speed of sound in air is 1,088 ft. per second. How many miles per hour is this?

187. If you can beat someone by 8 yards in a 200-yard sprint and he or she can beat someone by 6 yards in a 100-yard sprint, by how much would you beat the loser of the 100-yard race in a 50-yard sprint?

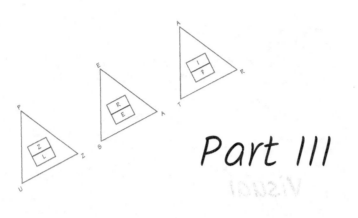

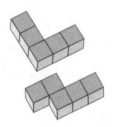

Part III

MATHEMATICAL REASONING

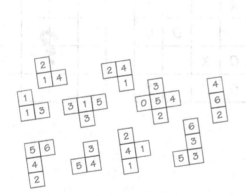

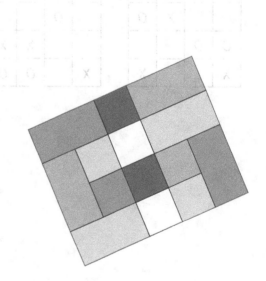

188. Below are three grids that contain X's and O's. See if you can determine the logic of the placement of the letters to produce the fourth grid.

X		X	O
		X	O
O	O		
X			X

X	O		X
	O		
		X	X
X		O	O

X			X
		O	O
O	X		
O	X		X

	?		

189. Three of the four shapes below can be made by assembling the two pieces at right. Which shape below is the odd one out?

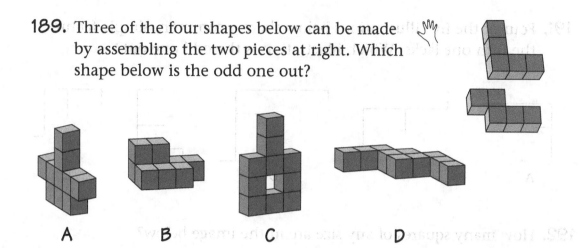

A   B   C   D

190. How many blocks are in the illustration below? Consider that all rows and columns run to completion unless you actually see them end.

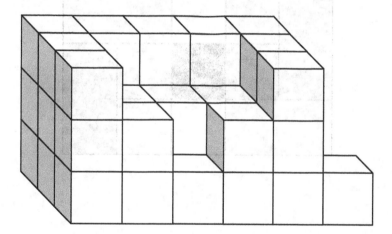

191. Four of the five illustrations below share a common, simple feature the fifth one lacks. Which illustration is the odd one out?

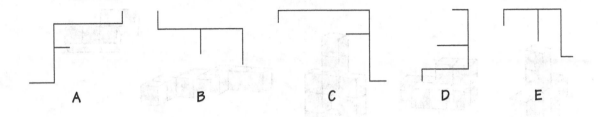

A   B   C   D   E

192. How many squares of any size are in the image below?

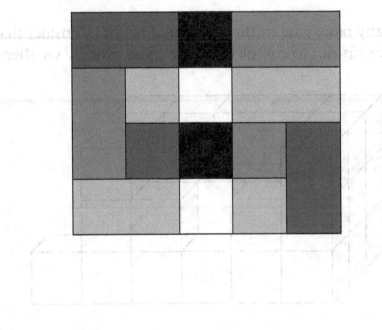

193. What is the total number of individual cubes in the two stacks below? Consider all rows and columns running to completion unless you actually see them end. Be careful with this puzzle!

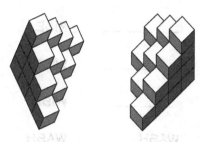

194. Here's a puzzle in analytical reasoning. Each symbol is represented by a letter, as are the number of symbols in each figure and how they are positioned with each other. See if you can figure out the code and come up with the two answers for the question marks.

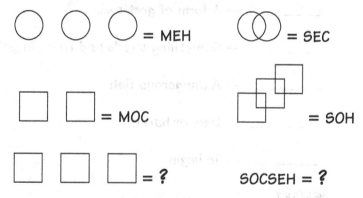

195. Here's a puzzle called a Trickledown. The goal is to change one letter on each line to arrive at the final word. Here is an example.

RIDE   RIDE

_____   RISE

_____   WISE

_____   WISH

WASH   WASH

Try this Trickledown. I'll give you a hint on each line.

THINK

_____ → A form of gratitude

_____ → Something that's bad to do in golf

_____ → A dangerous fish

_____ → Bare or harsh

_____ → To begin

SMART

Copyright © 2009 by John Wiley & Sons, Inc.

Just for Fun: Frame Game

196. Find the hidden word or phrase.

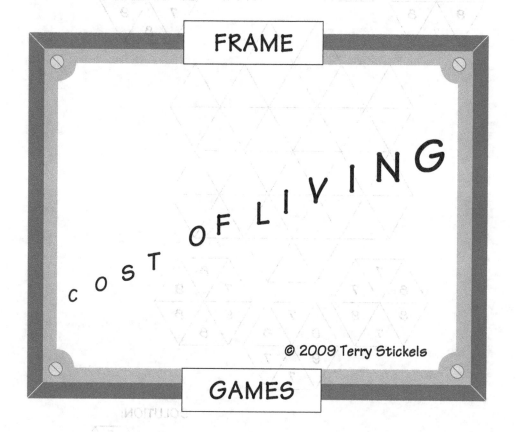

FRAME

COST OF LIVING

© 2009 Terry Stickels

GAMES

197. Can you place the hexagons into the grid so that where any hexagon touches another along a straight line, the numbers match? No rotation of any hexagon is allowed.

Here is an example:

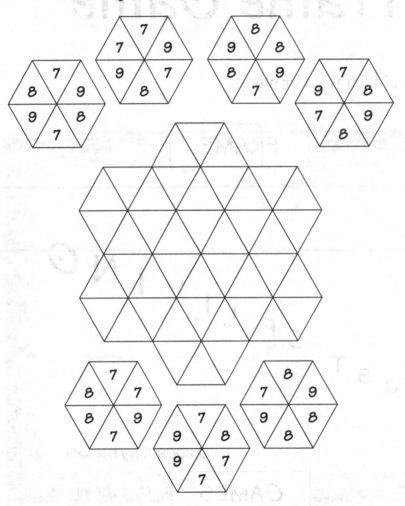

SOLUTION:

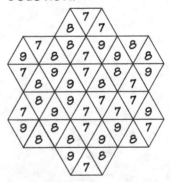

198. These puzzles lend themselves to fun group activity as well as solving individually. Pair up into teams and see how long you take to find your way through the maze below:

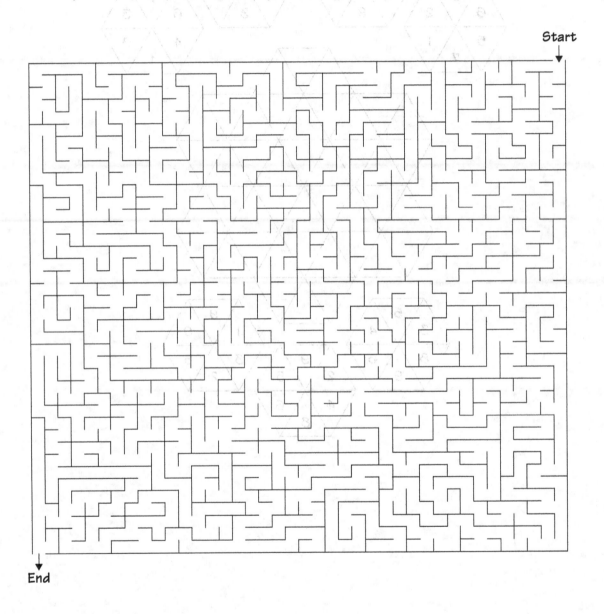

Start

End

199. Here is a 5 × 5 Magic Square. Using the numbers 1–25, arrange them so that every row and column add up to 65. Diagonal rows count as well. We have filled in a few numbers for you.

21	3	10		19
			1	
				22
18			9	
7		16		5

200. Four symbols are placed in pairs on five cards as follows:

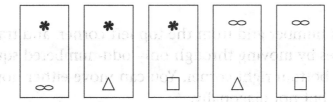

Here's the puzzle: What two symbols should be on the sixth and final card of the series?

201. This puzzle is called a zigzag puzzle. The goal is to enter the grid from the number 1 in the top left corner, and to go through all the numbers, in order, once and only once. The path cannot cross itself, but it can go horizontally, vertically, or diagonally.

Start

1	2	4	1	2	2	3	4
3	3	1	2	1	3	1	1
4	2	4	3	4	4	3	2
1	3	2	4	1	2	1	3
1	2	1	4	4	3	2	4
2	4	3	3	1	2	1	3
4	3	3	4	2	1	4	3
1	2	1	2	3	4	2	4

End

202. Enter this number grid from the top left corner, and travel through the squares by moving through only odd-numbered squares to reach the bottom right corner. You can move either horizontally or vertically, but not diagonally.

Start

3	9	7	5	3	7	3	6	7
7	2	9	8	6	2	5	8	3
5	8	5	2	3	6	3	7	9
3	6	7	4	7	4	2	6	3
7	4	3	9	5	3	5	2	7
9	2	9	6	8	6	4	6	8
3	6	7	3	5	2	7	5	9
5	4	2	8	3	8	3	6	7
7	3	5	6	9	5	7	2	9

End

203. Illustration 1 is a square piece of paper, Illustration 2 is the piece of paper folded in half, and Illustration 3 is the piece of paper folded in fourths.

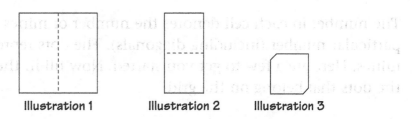

Illustration 1   Illustration 2   Illustration 3

Imagine that you have snipped off the opposite corners as shown in Illustration 3.

Now open the piece of paper. The result will look like:

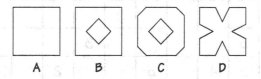

A   B   C   D

204. If you wanted to buy a new Harry Potter book, you had to hustle to the store to get in line.

$$\begin{array}{r} POTTER \\ -HARRY \\ \hline HURRY \end{array}$$

Above is an alphametic where each letter stands for a digit between 0 and 9. Zero cannot begin a word. Let A = 6, E = 9, and U = 2. What values for the other letters fit into this equation?

205. Logic minesweeper games are based on the famous computer game Minesweeper. The puzzle version does not rely on luck but, instead, pure logic to solve the puzzle.

The number in each cell denotes the number of mines around that particular number (including diagonals). The dots represent the mines. Here are a few to get you started. Now fill in the rest of the dots that belong on the grid.

1		0			1			2	
●			2	●		1			
●	●	●		1	1		1	2	
2		3		1		1		2	
2		2	●			3	3		
				2					2
3				3			3		
3	4		3		2		3		
			3				3		
3			1		1		1		

Just for Fun: Frame Game

206. Find the hidden word or phrase.

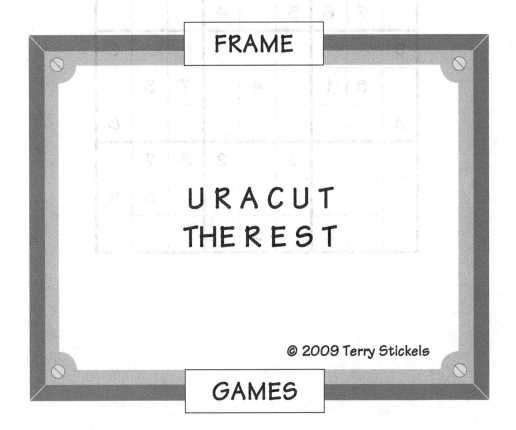

FRAME

URACUT
THEREST

© 2009 Terry Stickels

GAMES

207. Here is a Sudoku puzzle. Fill in the gameboard so the numbers 1 through 9 occur exactly once in each row, column, and 3 × 3 box. The numbers can appear in any order, and diagonals are not considered.

				6		8		
8	9		3					
	7	6	5		4			
9								2
	5	1		4		7	3	
3								8
			4		2	6	7	
					1		8	9
		2		9				

208. This next puzzle is called Killer Sudoku. The rules for Killer Sudoku are simple. The rules for regular Sudoku apply, with one additional rule: The sum of the cells in a cage (a group of cells surrounded by a dashed line) must equal the total given for the cage. Each digit in the cage must be unique.

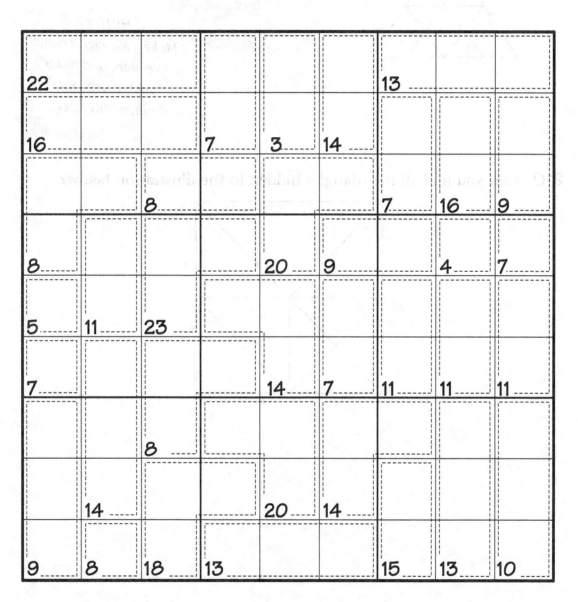

209. How many triangles of any size or shape are in the figure below?

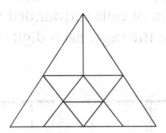

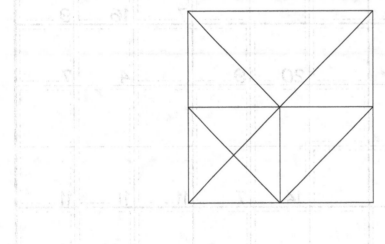

HINT #1
Think of an orderly way to go about this.

HINT #2
Make your own smaller version, and letter each place where lines intersect.

210. Can you find all the triangles hidden in the illustration below?

211. This puzzle has been turned into a jigsaw puzzle! The completed grid has been broken into smaller number-pieces below the grid. Put the number-pieces back into the grid so every row and column equals 12.

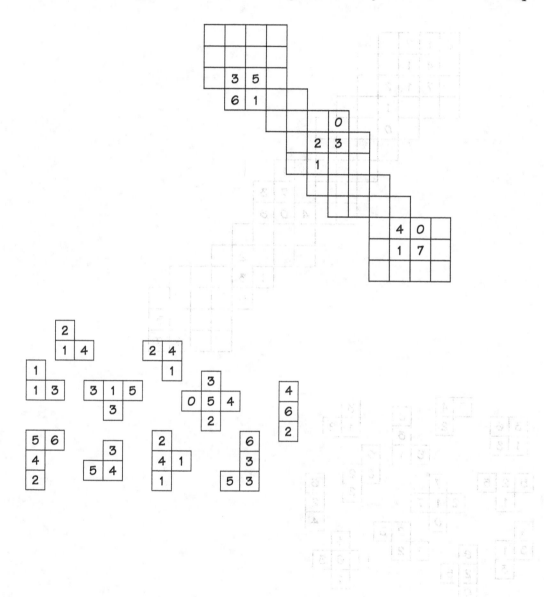

212. Follow the instructions for the previous puzzle, but this time put the number pieces back into the grid so that every row and column equals 15.

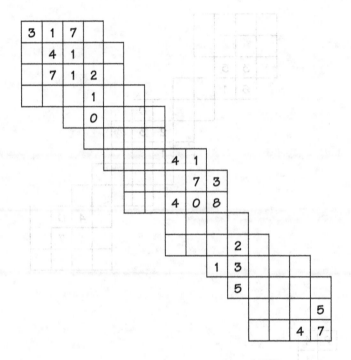

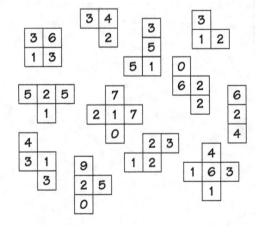

213. Here's a challenge in three dimensions! Can you figure out how many blocks are in each il lustration?

HINT
Hidden blocks follow the same patterns as the visible blocks.

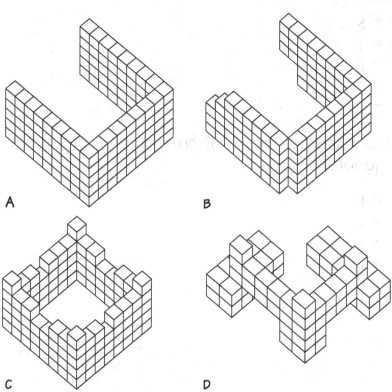

A

B

C

D

214. Use the clues given to solve the following cryptograms and find quotes from some famous characters.

a. [cryptogram in symbols]

◇ = S
◆ = T
▢ = R
◼ = O
Ж = I

b. Νεϖερ λεαϖε ψουρ φοοδ δισῃ υνδερ α βιρδ χαγε.
   —γαρφιελδ

ρ = R
λ = L
φ = F
γ = G

Just for Fun: Frame Game

215. Find the hidden word or phrase.

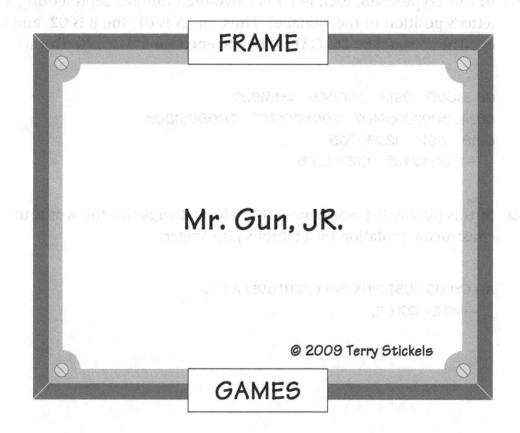

FRAME

Mr. Gun, JR.

© 2009 Terry Stickels

GAMES

216. To solve this cryptogram, you have to unscramble the individual words.

I KNTHI HATT YAEBM FI MOWEN DNA
LDICHNER REWE NI AGERCH, EW UODLW
ETG MESOEREWH.
　—SEMAJ URBRETH

217. In this cryptogram, each letter is a two-digit number representing a letter's position in the alphabet. Thus, the A is 01, the B is 02, and so on. The Z would be 26. CAT would be encoded as 030120. Got it?

08151605　0919　200805　13151920
0524030920091407　2008091407　2008051805
0919　0914　12090605.
　—1301140425　1315151805

218. In this puzzle, the words are out of place. Reorganize the words to construct a quotation by a famous pop singer.

AN CYRUS JUST PINK ISN'T ATTITUDE! A IT'S
　—MILEY COLOR,

219. The answer to this puzzle is taken from an actual fortune cookie. Answer the clues below to find each word. Then place the letters on the lines at the bottom of the page. Each letter must be placed on a line marked by the same picture. When all the lines are full, your fortune will be revealed.

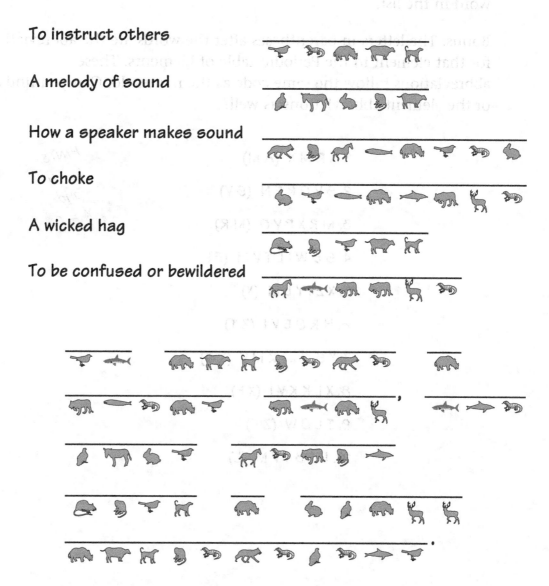

To instruct others

A melody of sound

How a speaker makes sound

To choke

A wicked hag

To be confused or bewildered

220. Here's a new kind of coded puzzle for you to try: a cryptolist! The word list below contains natural elements found on the Periodic Table of Elements—like Sodium (Na) or Potassium (K). The same code is used for every element. For example, if you determine that the letter A is really the letter E in the first word, then it is true for every word in the list.

Bonus: The letters in parentheses after the words are the abbreviation for that element in the Periodic Table of Elements. These abbreviations follow the same code as the names do. Can you find all of the element abbreviations as well?

HINTS
S = H
and
X = C

1. A R M X (A M)

2. S V O R F N (S V)

3. M R X P V O (M R)

4. S B W I L T V M (S)

5. X Z I Y L M (X)

6. H R O E V I (Z T)

7. X S O L I R M V (X O)

8. X L K K V I (X F)

9. T L O W (Z F)

10. L C B T V M (L)

221. When the illustration below is folded into a cube, the side opposite the W is _?_.

When the W is positioned as it is here, the figure opposite the W looks like _?_.

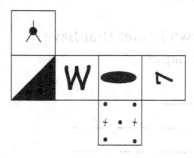

222. Based on the logic used for the first two boxes, what four different positive digits will go in the final box?

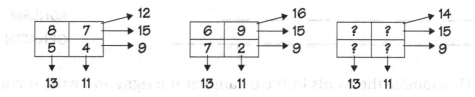

223. Can you cut this 2 × 12 rectangle into two congruent pieces . . .

. . . and place the two pieces together to form this 3 × 8 rectangle?

224. Unscramble the words in the column at the right and write them correctly in the spaces provided. The first letter of each unscrambled word, reading from top to bottom, spells out the answer to this question:

HINT
All of the unscrambled words in this and the following two puzzles relate to math.

What do you call two figures that have the same size and shape?

___ ___ ___ ___   EBUC

___ ___ ___ ___ ___ ___   BTSOUE

___ ___ ___ ___ ___ ___   UMERBN

___ ___ ___ ___ ___   RAHPG

___ ___ ___ ___ ___ ___ ___ ___   ENMDAIERR

___ ___ ___ ___   TNIU

___ ___ ___ ___ ___ ___ ___   OXETNPEN

___ ___ ___ ___ ___ ___   RONLAM

___ ___ ___ ___ ___ ___ ___   GENATTN

225. Unscramble the words in the column at the right and write them correctly in the spaces provided. The first letter of each unscrambled word, reading from top to bottom, spells out the answer to this question:

What is a closed-plane figure formed by three or more line segments that do not cross over each other?

___ ___ ___ ___ ___   RMIEP

___ ___ ___ ___   VALO

___ ___ ___ ___   ENIL

___ ___ ___ ___ ___ ___ ___ ___ ___   ITSDRAYCK

___ ___ ___ ___ ___ ___ ___   ETYRMOGE

___ ___ ___ ___ ___ ___   UTUTOP

___ ___ ___ ___ ___ ___   REBUNM

226. Unscramble the words in the column at the right and write them correctly in the spaces provided. The first letter of each unscrambled word, reading from top to bottom, spells out the answer to this question.

What is the basic arithmetic operation symbolized by a minus sign?

__ __ __ __ __ __   QUASER

__ __ __ __ __   NOIUN

__ __ __ __ __ __   ECTBIS

__ __ __ __ __ __ __   AGNETNT

__ __ __ __   LEAR

__ __ __ __ __ __ __ __   SOBAULET

__ __ __ __ __ __ __ __   STONCNTA

__ __ __ __ __   OTTLA

__ __ __ __ __ __ __ __ __   SCSOILEES

__ __ __ __ __ __   BLOGNO

__ __ __ __ __ __ __   TNAALRU

227. Unscramble the words in the column at the right and write them correctly in the spaces provided. The first letter of each unscrambled word, reading from top to bottom, spells out the answer to this question:

What is the measure of the number of cubic units needed to fill the space inside an object?

__ __ __ __ __ __ __ __   TICLAVER

__ __ __ __ __   CETTO

__ __ __ __ __   EELVL

__ __ __ __   TUIN

__ __ __ __ __ __   XAMIRT

__ __ __ __ __ __ __ __   ITAQEUNO

Just for Fun: Frame Game

228. Find the hidden word or phrase.

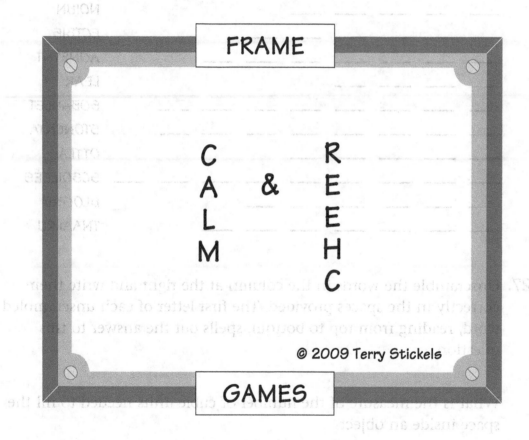

FRAME

C A L M & R E E H C

GAMES

© 2009 Terry Stickels

229. The King had a large plot of land that he wanted to leave to 7 of his Princes. It was a square piece of property with 7 gold mines on it. The King really liked puzzles and decided to give all 7 gold mines to the first Prince who could divide the land with three lines, and three lines only, that divided the land with one gold mine in each section. That Prince would control all 7 mines. Here's how the mines were situated.

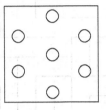

Months went by and no one could solve the puzzle until one day a poor young farmhand wrote a letter to the King with a drawing of how it could be done. The King was so impressed that he gave the land—and the hand of the King's beautiful daughter, the Princess of the kingdom—to the young farmhand. How did he draw the lines so there was one mine in each of the 7 sections?

Note: The 7 sections do not have to be equal in area.

230. How many faces or sides does this illustration have?

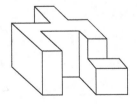

231. Start

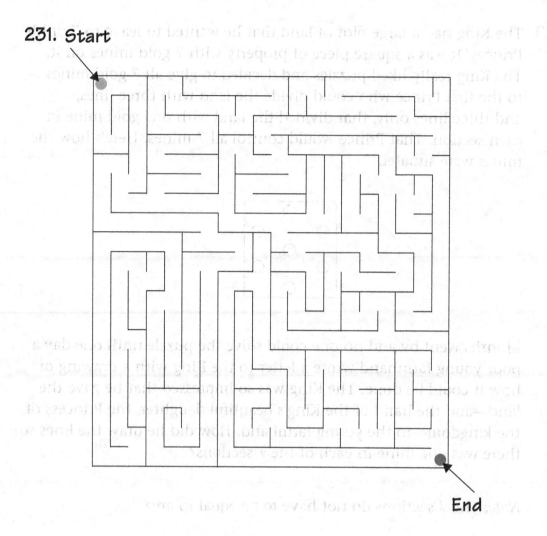

End

232. Can you retrace the illustration below by starting at "A" and ending at "B" without retracing any line already drawn or crossing any lines?

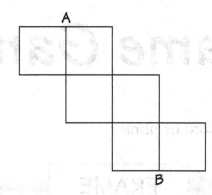

233. Four of the images below can be created without lifting your pencil, or intersecting or retracing any lines. Which illustration is the odd one out?

1   2   3   4   5

234. What comes next in this sequence?

₿ ℃ ⅅ Ɛ Ϝ ǤǤ ⱧⱧ ‖

a. #

b. ∝

c. ⨆

d. AA

Just for Fun: Frame Game

235. Find the hidden word or phrase.

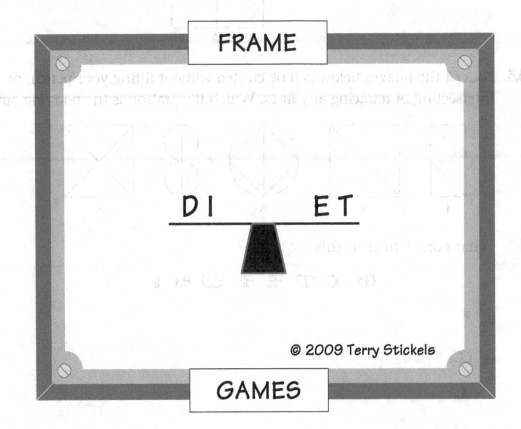

FRAME

DI   ET

© 2009 Terry Stickels

GAMES

236. Can you draw two squares inside the square below so each goldfish has its own separate area to swim in?

237. The letters in the five squares are positioned in a logical, sequential manner so they will fall into place in the blanks below the squares. The sentence has 25 letters and starts with an "S" and ends with an "R."

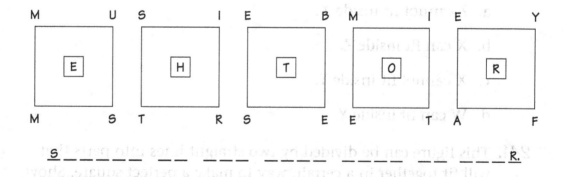

S _ R.

238. How many squares of any size can be found in this illustration?

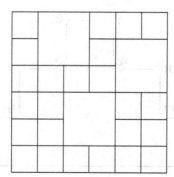

239. The letters in and around the triangles are placed in a logical manner so they spell out a simple sentence. What is the missing letter in the last triangle, and what is the sentence?

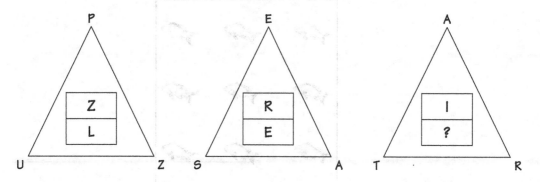

240. There are four boxes labeled W, X, Y, and Z. W and Z are the same size. W can fit inside X, and Y can fit inside Z.

Knowing this, which one of the following statements is true for certain?

a. Z cannot fit inside X.

b. X can fit inside Z.

c. X cannot fit inside Y.

d. W can fit inside Y.

241. This figure can be divided by two straight lines into parts that will fit together in a certain way to make a perfect square. Show where the two lines have to be drawn so those two parts can be moved and placed on the figure to make a square. All angles are 90°.

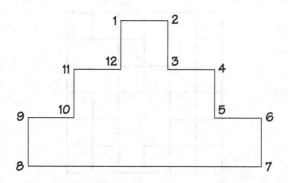

242. How many times do the sides or faces only (no edges) of a block touch the sides or faces of other blocks? Each block has its own count toward the total. Each block is the same size and has six total faces or sides. *Do not count edges that touch!*

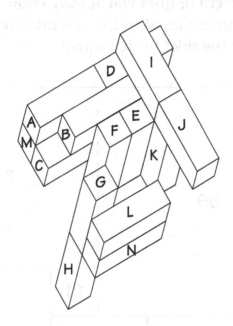

Example: Block A touches B, D and M.

Example: Block J touches I and K.

Example: Block C touches M, B, and F.

243. In this puzzle, called a "squared square," squares of different sizes are contained within one big square. This puzzle has many versions. Some, involving rectangles and triangles, are equally as fun. The goal is to find out the sizes of the squares with the question marks. By comparing length of lines you already know, you can make some deductions to find out the sizes that are missing. Each number stands for the length of the sides in that square.

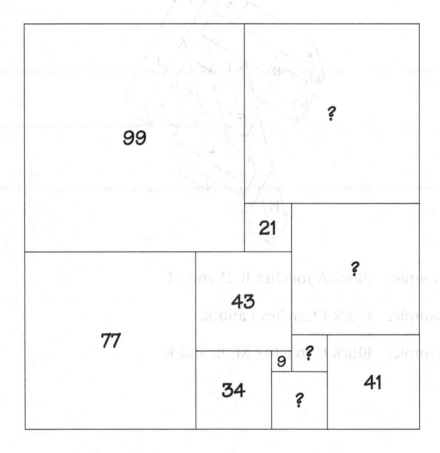

Just for Fun:
Frame Game

244. Find the hidden word or phrase.

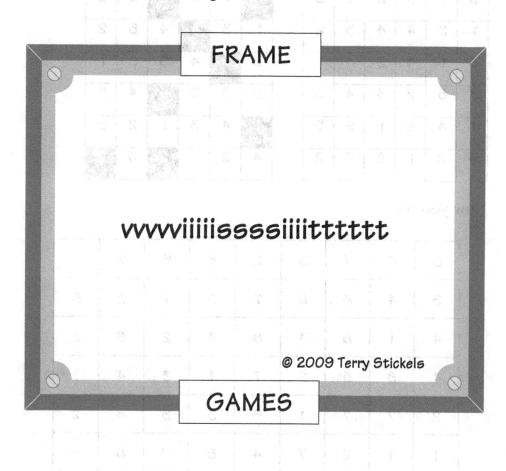

FRAME

wwwiiiiisssssiiiittttttt

© 2009 Terry Stickels

GAMES

245. Hitori is a number puzzle played on a grid. The object is to black out squares so that duplicate numbers do not appear in a row or column more than once. Blackened squares must not touch each other vertically or horizontally. The squares that are white must form a single continuous area. This means they must touch side-to-side or corner-to-corner.

Example:

5	1	5	2	2	3
1	3	4	4	5	2
5	1	4	3	4	1
3	5	2	4	4	5
1	4	3	1	2	5
4	2	1	2	3	2

5	1	5	2	2	3
1	3	4	4	5	2
5	1	4	3	5	1
3	5	2	4	4	5
1	4	3	1	2	5
4	2	1	2	3	2

Now you try.

5	6	7	3	2	8	5	7	1
3	4	8	6	7	6	1	2	5
4	1	8	4	8	4	2	6	3
4	8	5	2	7	1	3	4	7
2	7	7	1	5	3	3	4	2
1	1	2	7	4	5	6	8	4
2	5	3	1	6	2	7	1	8
3	2	5	8	1	4	5	5	6
5	3	7	6	1	3	4	2	3

246. Based on the logic of the first nine boxes below, what should the next three boxes look like?

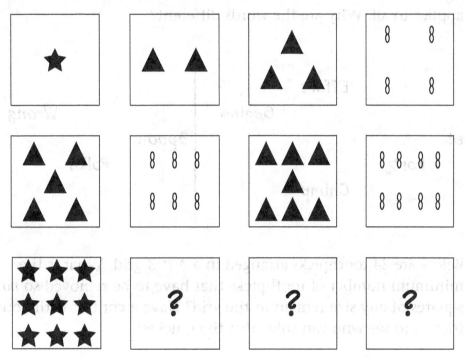

247. Below is the top view of a stack of cubes. The numbers in each cell represent the number of individual cubes in that column. Which lettered stack is represented by the numbers in the cells?

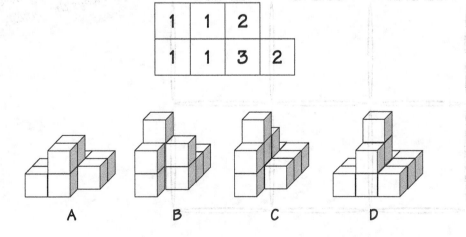

248. The words on the left of the line are different from the words on the right based on construction of the words. A straightforward reason applies to all. Why are the words different?

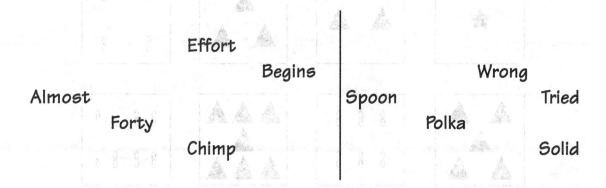

Effort

Begins   Wrong

Almost   Spoon   Tried

Forty   Polka

Chimp   Solid

249. Below are 24 toothpicks arranged in a 3 × 3 grid. What is the minimum number of toothpicks that have to be removed so *no* squares of any size remain in the grid? Have a contest with your friends to see who can solve this the quickest.

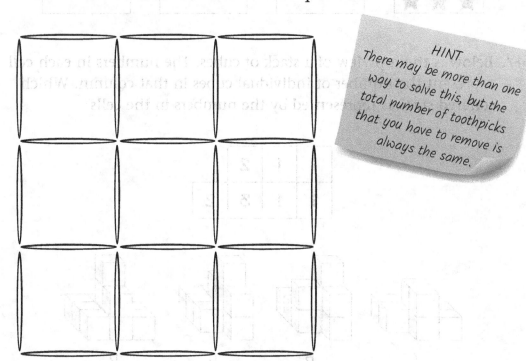

250. A certain number of identical sheets of paper are placed overlapping, as shown below. How many sheets of paper were used to make this illustration?

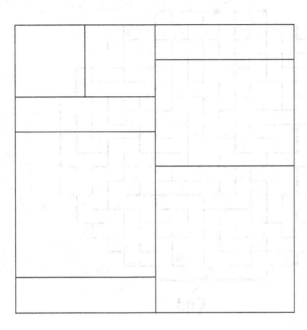

251. One of the illustrations below does not belong with the others, based on a simple design feature. Which is the odd one out? Why?

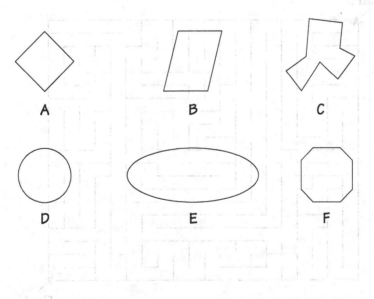

252. Find your way through these two mazes.

a.

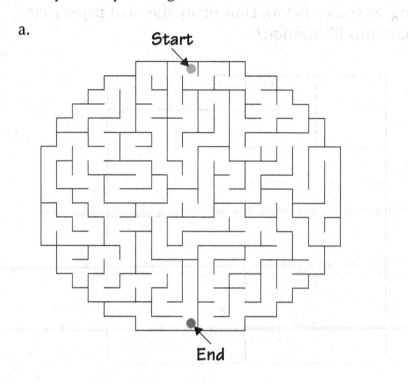

b.

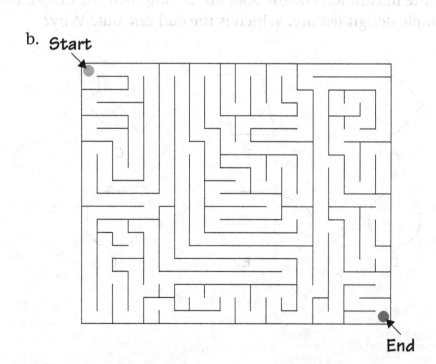

Just for Fun: Frame Game

253. Find the hidden word or phrase.

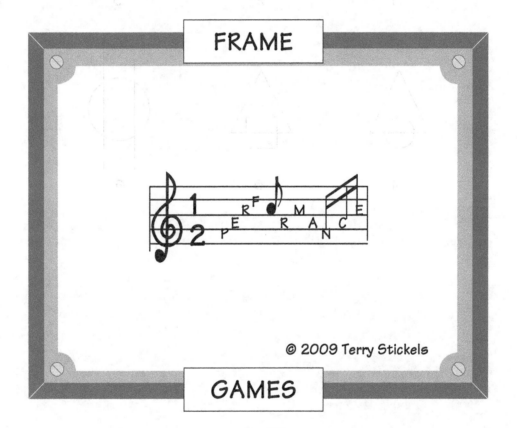

254. One of the six figures below doesn't belong with the other five figures, based on a simple design feature. Which is the odd one out?

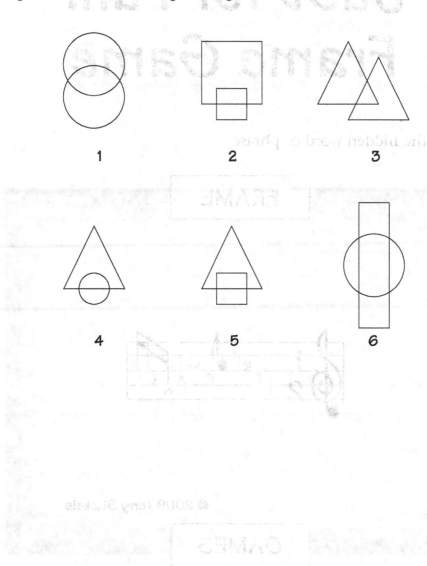

255. A newspaper delivery service has several routes throughout the city. The service tries to pick routes that are the most efficient and end up back where the delivery started. Here's an example:

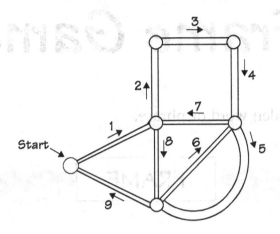

By following the numbers from 1–9, you can see that the route ends back where it started without backtracking on any streets twice.

Sometimes getting back to the starting point isn't possible without going down at least one street twice, but it is possible to complete the route at a different location.

One of the following routes can be completed without retracing, but it will finish in a location different from where it started. The other route cannot be completed without going on at least one street twice. Which is which?

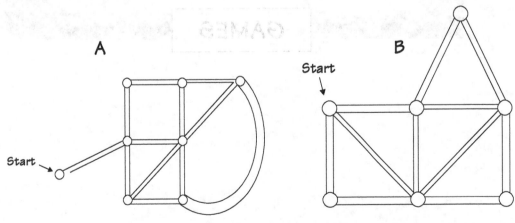

Just for Fun:
Frame Game

256. Find the hidden word or phrase.

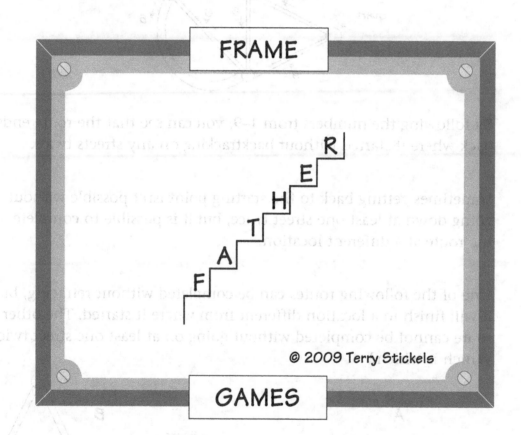

FRAME

FATHER

© 2009 Terry Stickels

GAMES

257. Kareem and Charles were talking about an old puzzle where you put 10 pennies in a circle as pictured below:

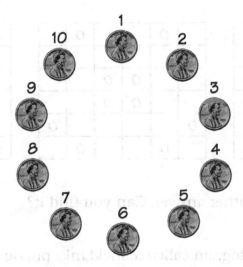

Starting at number 1 and counting by threes, you eliminate every third coin. The puzzle asks which coin will be the last one remaining. The answer is number 4. Kareem says to Charles, "That means if I count by fives instead of by threes, the last coin remaining will be coin number 6."

Charles said, "You're way off. The last coin standing is coin number:

a. 10

b. 7

c. 3

d. 1

258. An old puzzle asks you to place 12 circles on a 6 × 6 grid so each row, column, and diagonal contains exactly two circles. Here are the answers that are usually given.

		O	O		
	O			O	
O					O
O					O
	O			O	
		O	O		

	O			O	
O					O
		O	O		
		O	O		
O					O
	O			O	

O					O
		O	O		
	O			O	
	O			O	
		O	O		
O					O

But there is another answer. Can you find it?

259. Below is a cryptogram called a StickLinks puzzle. Starting with the letter "I" and finishing with "You," link the letters together one by one—horizontally, vertically, diagonally, and backward and forward—to make words. (There are two sentences.)

```
(I)  O   E   G   T   H   T   F   U
 H   P   Y   E   S   E   K   N   M
 E   U   O   P   Z   L   I   A   T
 N   O   E   U   Z   E   N   G   H
 J   Y   D   H   I   S   F   M   E
 L   O   S   A   G   E   O   R  (Y
 V   I   N   D   R   A           O
                                 U)
```

— Terry Stickels

— ———— ——— ———————— ———————— —————— ————————.
— ——— —————— ——— ———————— ————— ———— ————.

Other

260. If Sara is twice as old as Eileen will be when Gina is as old as Sara is now, who is the oldest, the next oldest, and the youngest?

261. The letters below are separated in this manner for a specific reason. What is it?

Q	W		O	P
A	S		K	L
Z	X		N	M

HINT
Try typing these letters.

262. Find the math words listed below in this word search puzzle.

```
N  M  A  W  Z  U  F  H  U  G  P  N  G  Y  Z
Y  O  L  E  L  L  A  R  A  P  O  M  E  T  E
T  E  I  M  E  E  R  E  P  G  M  S  O  I  R
A  P  H  T  V  L  Q  O  A  O  N  V  M  N  O
N  O  E  V  A  U  G  X  T  O  W  Y  E  I  S
G  L  J  X  A  C  E  N  I  C  C  E  T  F  A
E  S  R  T  P  H  I  T  A  K  A  F  R  N  R
N  J  I  T  A  O  C  L  G  I  W  F  Y  I  O
T  O  R  B  U  A  N  D  P  L  R  R  G  R  G
N  T  B  E  R  Q  A  E  E  I  D  T  A  T  A
W  W  M  F  J  J  X  C  N  C  T  D  S  D  H
R  E  C  T  A  N  G  L  E  T  I  L  E  R  T
I  N  T  E  G  E  R  D  R  C  K  M  U  V  Y
A  R  B  E  G  L  A  K  A  Y  P  G  A  M  P
Y  X  A  H  J  E  X  L  C  U  B  E  S  L  R
```

ALGEBRA	CUBES	DECIMAL
EQUATION	EXPONENT	FACTOR
FRACTIONS	GEOMETRY	HEXAGON
INFINITY	INTEGER	MULTIPLICATION
PARALLEL	POWER	PYTHAGORAS
RADICAL	RECTANGLE	SLOPE
TANGENT	TRIANGLE	ZERO

263. Thirteen pennies are put into three piles so each pile has a different number of pennies. This can happen in eight different ways. Here are two:

1, 2, 10

1, 3, 9

Can you find the other six ways?

264. A jogger trains by running through several small towns. He runs two different routes from Alser to Bahnsen. He knows three different routes from Bahnsen to Corning. From Corning to Denton, he knows five different running paths. How many different paths does he know from Alser to Denton?

Just for Fun: Frame Game

265. Find the hidden word or phrase.

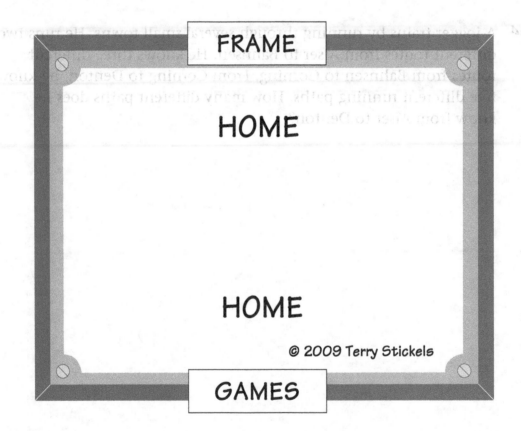

FRAME

HOME

HOME

© 2009 Terry Stickels

GAMES

266. In a foreign language, "Kar Nola Roz" means "move four apples." "Kir Roz Pala" means "sell four plates." "Insa Kar Kir" means "carefully move plates." Which word means "apples"?

267. If box A can fit inside box B, box C can fit into box D, and boxes B and C are the same size, which of the following must be true?

  a. C will fit into A.

  b. A could be larger than D.

  c. B and C are twice as large as A.

  d. B will fit into D.

268. In the box at the right is a scrambled 14-letter word that is used in common everyday English. See how long it takes you to unscramble it.

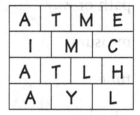

A	T	M	E
I	M	C	
A	T	L	H
A	Y	L	

269. A driver notices that her odometer reads 16961—a palindrome (something that reads the same backward as it does forward). What is the next mileage reading that will be a palindrome?

270. Mark's father saw that his odometer read 82,222. Mark asked his dad, "The number 82,222 has four of the same digits. What is the next mileage reading that will have four of the same digits?" Can you help Mark's dad?

271. "Odo" is a Greek prefix from the Greek word *hodos*, which (used in mathematics) means:

    a. miles

    b. path or way

    c. measure

    d. time

    e. wheels

Just for Fun: Frame Game

272. Find the hidden word or phrase.

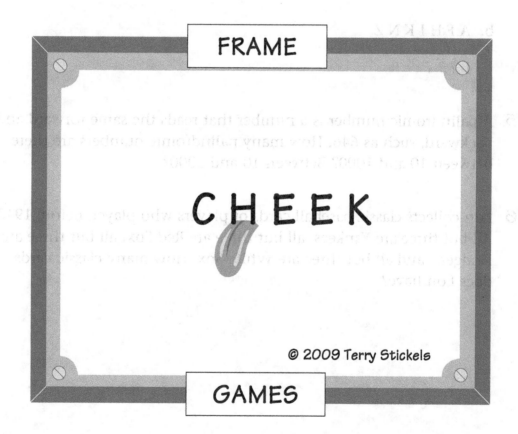

FRAME

CHEEK

© 2009 Terry Stickels

GAMES

273. How many different words of any length can you find in the word "quarter"? We've found 21—and there may be more!

274. The capital letters below are grouped because of a certain characteristic they share. See if you can determine why the letters were grouped into three different groupings and then come up with the last three letters in group c.

a. E M W

b. A F H I K N Z

c. L ? ? ?

275. A palindromic number is a number that reads the same forward and backward, such as 646. How many palindromic numbers are there between 10 and 1000? Between 10 and 2000?

276. Lon collects classic baseball cards of players who played before 1940. All but three are Yankees, all but three are Red Sox, all but three are Dodgers, and all but three are White Sox. How many classic cards does Lon have?

277. Four friends, Barbara, Bianca, Patty, and Tina, are nicknamed Reelo, Wheezie, Fly, and Blaze—but not necessarily in that order. Based on the clues below, what is the nickname of each friend?

HINT
Make a chart and eliminate possibilities.

a. Patty can run faster than Reelo but can't swim as fast as Fly.

b. Reelo is a faster swimmer than Tina but can't run as fast as Wheezie.

c. Barbara is faster than both Patty and Blaze but can't swim as fast as Reelo.

278. A local ice cream shop is having a special on sundaes. You can get two different flavors of ice cream in a large sundae for the price of one scoop. The featured flavors are:

Cherry   Mint

Vanilla   Mocha

Chocolate   Strawberry

Jack and his sister, Lori, wondered how many different combinations of two scoops each a person could order. Can you help them out?

Just for Fun: Frame Game

279. Find the hidden word or phrase.

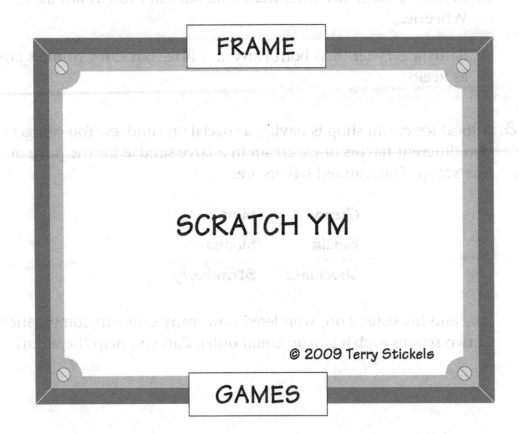

FRAME

SCRATCH YM

© 2009 Terry Stickels

GAMES

280. I have two containers. One is 7 gallons, and one is 9 gallons. I need exactly 1 gallon of water from one of these containers but have no measuring devices. I have access to all the water I need, so the containers can be filled any way I choose. Is it possible to pour water back and forth between these containers to come up with exactly 1 gallon? If so, how can it be done? Remember—I can fill the containers at will and transfer the water back and forth between containers to come up with different amounts.

281. Snails move slower than ants, and snails are smaller than moles. Snails move faster than moles, and snails are larger than ants.

One of the following is true based on the statements above.

a. Ants are the fastest and second largest.

b. Snails are the second in size and slowest.

c. Moles are the slowest and largest.

282. The years below are called "upside downside" years; they read the same when turned over. I've given you the six consecutive years when this occurs. What is the next "upside downside" year?

1691   1961   6009   6119   6699   6969   ?

What is the year before 1691 when the year is an "upside downside" year?

HINT
Only 0, 1, 6, and 9 are used for the digits in the years.

283. Molly's father's uncle's son's cousin could be:

   a. Molly

   b. Molly's father

   c. Molly's grandfather

   d. Molly's sister

284.

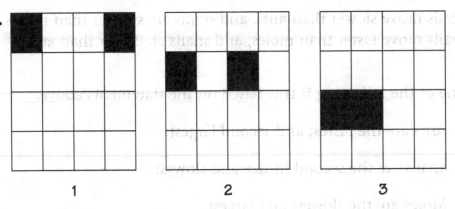

The next and final pattern of the grid above would look like which of the following:

HINT
Think of movement.

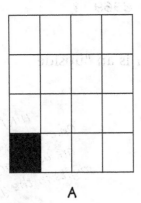      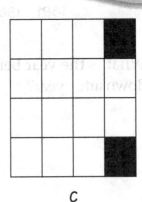

285. One of these words is different from the others. Which is the odd one out?

Madam   Drawer   Radar

Repaper   Rotator

Level

286. One of the objects below does not belong with the others, based on its function. Which is the odd one out?

a. umbrella

b. music stand

c. rain gauge

d. truck

e. magnet

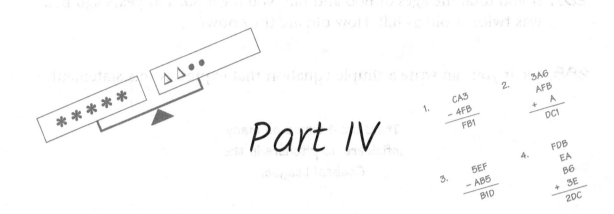

Part IV

ALGEBRA, STATISTICS, and PROBABILITY

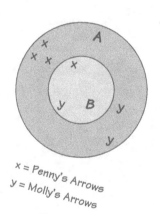

x = Penny's Arrows
y = Molly's Arrows

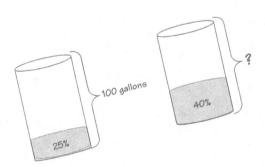

100 gallons

25%

40%

?

287. If you total the ages of Bob and Bill, you'll get 50. Ten years ago Bob was twice as old as Bill. How old are they now?

288. See if you can write a simple equation that expresses this statement:

There are 4 times as many
infielders as pitchers in the
Coastal League.

Use I for infielders and P for pitchers.

289. Seven times a number is four more than three times the number. What is the number?

290. I'm thinking of a number that when added to $1\frac{1}{2}$ will give me the same result as when it is multiplied by $1\frac{1}{2}$. What is that number?

Just for Fun: Frame Game

291. Find the hidden word or phrase.

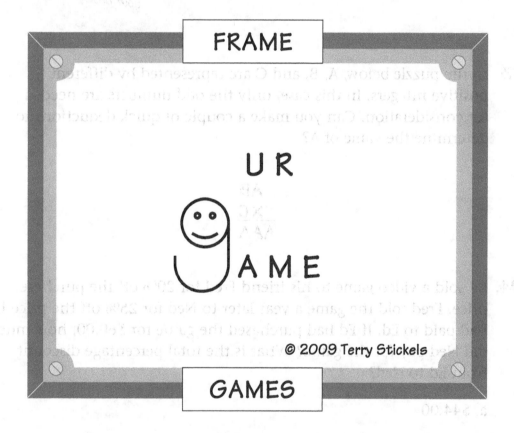

© 2009 Terry Stickels

292. I'm thinking of a two-digit number where adding the two digits together will give you $\frac{1}{3}$ of the original two-digit number. What is the number?

HINT
You probably could guess the number in short order. The puzzle becomes a little more challenging if you can come up with an equation for proving your answer.

293. In the puzzle below, A, B, and C are represented by different positive integers. In this case, only the odd numbers are needed for consideration. Can you make a couple of quick deductions to determine the value of A?

$$
\begin{array}{r}
AB \\
\times C \\
\hline
AAA
\end{array}
$$

294. Ed sold a video game to his friend Fred for 20% off the purchase price. Fred sold the game a year later to Ned for 25% off the price he had paid to Ed. If Ed had purchased the game for $60.00, how much did Ned pay for the game? What is the total percentage discount from Ed to Ned?

a. $44.00

b. $48.00

c. $36.00

d. $30.00

295. A + B = Z

Z + P = T

B + P = 8

What is the value of A? Choose from one of the three below.

$$A = 8 + P \qquad OR \qquad A = 8 \qquad OR \qquad A = T - 8$$

296. How many feet are in P yards and Q miles?

a. $3P + Q$

b. $5280 (P+Q)$

c. $3P + 5280Q$

297. Determine the missing value in the puzzle below.

$$\diamondsuit \bigstar = 8$$

$$\diamondsuit \diamondsuit \bigstar = 10$$

$$\diamondsuit \bigstar \bigstar \bigstar \bigstar = ?$$

298. If $4^n = 64$, what is the value of 2^{n+3}?

299.

$\triangle x = x^2 - 1$, if x is an even number

$\boxed{x} = x^2 + 1$, if x is an odd number

a. What is the value of $\triangle 6 - \boxed{5}$?

b. What is the value of $\triangle 8 + \boxed{3}$?

300. If $\dfrac{P}{Q} = \dfrac{4}{5}$, then $4Q \times 5P$ is ____? Consider P and Q as positive, whole numbers.

a. 20

b. a negative number

c. a square number

d. $\dfrac{5}{4} \times P$

301. A car can travel M miles in H hours. How many miles will the car travel in Y hours at the same rate?

a. $M \times H \times Y$

b. $\dfrac{M}{H} \times Y$

c. $\dfrac{M}{Y}$

d. $M + \dfrac{Y}{H}$

Just for Fun: Frame Game

302. Find the hidden word or phrase.

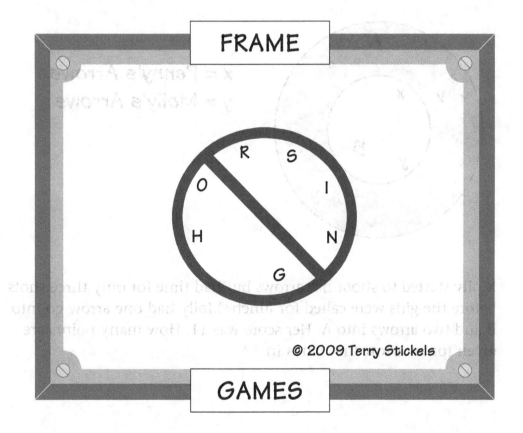

FRAME

R S
O   I
H   N
G

© 2009 Terry Stickels

GAMES

303. Here's a quick math brainteaser:

40% of 50 is 8% of __?__

304. Penny and Molly wanted to shoot arrows at a target on an outdoor range. Penny shot three arrows into the A target and one arrow into the B target, as shown below. Her total score for these four shots was 14.

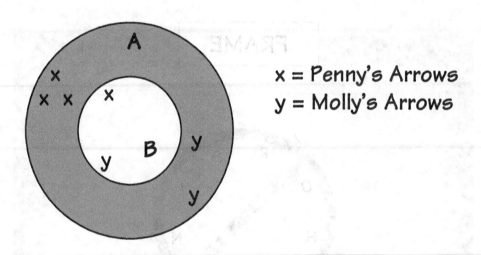

x = Penny's Arrows
y = Molly's Arrows

Molly started to shoot her arrows but had time for only three shots before the girls were called for lunch. Molly had one arrow go into B and two arrows into A. Her score was 11. How many points are given for an arrow that lands in A?

305. *What does a duck do when it flies upside down?* The answer is written below in code. To crack the code, solve these equations:

If I + 5 = 6, then I = ?   If 2U = S, then S = ?

If 2C = 12, then C = ?   If Q + 2 = 7 – T, then Q = ?

If 5 – A = 1 – I, then A = ?   If 3Q = P, then P = ?

If 2T + 2 = 6, then T = ?   If K – A + T = C – I – 1, then K = ?

If U – 3 = T – I, then U = ?

Riddle Answer: (substitute the number for the letter it equals)

1	2	3	4	5	6	7	8	4	9

306. If the following equations are true:

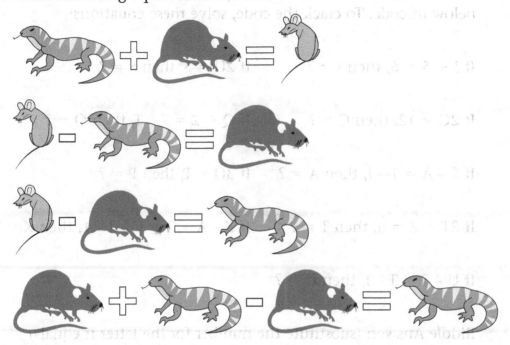

Then solve these:

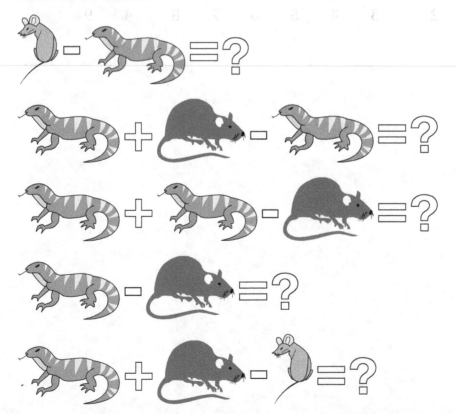

307. The equations below are partially in code. The numbers 1 to 6 have been replaced with letters from A to F (but not in that order). Can you crack the code and replace the letters with numbers to make the equations correct?

HINT #1
The same letter represents the same number in every equation.

For example, if you find that A = 3, then that's true for all four equations.

HINT #2
The equations use only the numbers 1, 2, 3, 4, 5, and 6.

1.
$$\begin{array}{r} CA3 \\ -\ 4FB \\ \hline FB1 \end{array}$$

2.
$$\begin{array}{r} 3A6 \\ AFB \\ +\quad A \\ \hline DC1 \end{array}$$

3.
$$\begin{array}{r} 5EF \\ -\ AB5 \\ \hline B1D \end{array}$$

4.
$$\begin{array}{r} FDB \\ EA \\ B6 \\ +\ 3E \\ \hline 2DC \end{array}$$

308. If A × B × C = 1 and B × C × D = 0, then:

    a. A is less than 1

    b. B is less than 1

    c. C = 1

    d. D = 0

    e. A = 0

309. A high school baseball team was going to its annual spring banquet when one of the ballplayers said, "Seven of us are outfielders, nine of us are infielders, and the rest are pitchers who make up one-third of the team." How many ballplayers are on the team?

310. Here's an alphametic puzzle paying tribute to baseball. Each letter stands for a positive integer from 0–9. Zero cannot begin a word. For this puzzle, let A = 4.

```
  FUN
  FUN
  FUN
 +FUN
 BALL
```

HINT

Don't forget—you might have to carry a number from one column to the next.

311. S and T represent whole numbers, and S Δ T means the same as $\dfrac{S + T}{3}$. What is the value of:

a. $5 \Delta (5 \Delta 7)$?

b. $\dfrac{6 \Delta 3}{7 \Delta 8}$?

312. x and y are two different numbers taken from the numbers $1-100$. What is the largest value that $\dfrac{x - y}{x + y}$ can have? What is the largest value that $\dfrac{x + y}{x - y}$ can have?

Just for Fun: Frame Game

313. Find the hidden word or phrase.

FRAME

B u c k **e** t

© 2009 Terry Stickels

GAMES

314. In an adult evening math class, the average age of the class of men and women is 36. The average age of the men is 45, and the average age of the women is 30. What is the ratio of the women to the men in the group?

315.
```
   YY
 +YY
  XYZ
```

If YY is a two-digit, positive number, what is the only value that Z could have?

316. The three scales below are perfectly balanced. If ● = 3, what are the values of Δ and ✳ ?

a.
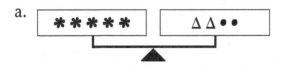

b.
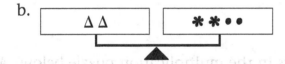

c.
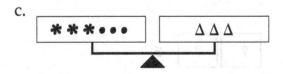

317. The product of two numbers is 168, and their difference is 2. What are the two numbers?

318. If $2 = \dfrac{1}{5}$, then what does 7 equal?

319. Annie has nickels and dimes in her pocket. She has three times as many dimes as nickels. Her friend Brenda says, "I can write an expression that will tell me exactly the amount of money you have in your pocket when you have three times as many dimes as nickels. All I need to know is how many nickels you have, and I can show you the correct amount."

If *x* equals the number of nickels, then Brenda's expression reads:

a. $x + 3x$

b. $35x$

c. $x(2x + .05)$

d. There is no expression.

320. What are the missing digits in the multiplication puzzle below? All the missing digits are different numbers.

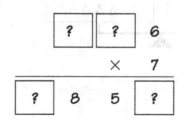

321. In the alphametic below, let A = 7, D = 5, M = 9, and T = 4. No word can begin with zero. What are the values of the other letters that will make this alphametic add up correctly?

```
      ADD
       IT
       UP
     MATH
   +   IS
   -------
    SHARP
```

322. A large carton of ice cream is balanced on a scale by a $\frac{5}{8}$ carton of the same ice cream plus $\frac{5}{8}$ of a pound. How much does the whole carton weigh?

323. Bill has 50 pieces of candy. There are 10 each of 5 different colors. The candy dish is on top of the refrigerator, so he can't see what colors he pulls from the dish.

His sister Linda says, "I'll bet you can't tell me how many pieces of candy you would have to take out of the dish to make sure you had five pieces of the same color."

Bill responded immediately with the correct answer. Can you?

324. You have a bag with six black cubes and three white cubes. Without looking when you make your selections, what is the probability of choosing two black cubes in a row?

325. A target of 6-inch diameter has a 3-inch center diameter. What is the probability that a random throw that hits the target will hit the center?

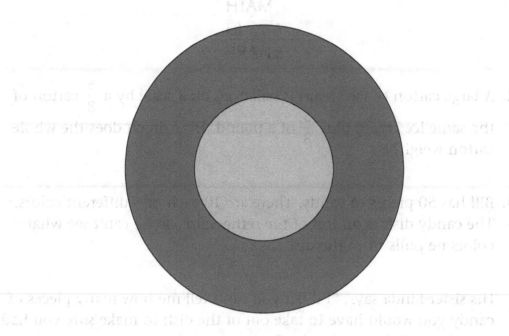

326. Here's a new twist to a puzzle. I'll give you the question and answer here:

If 40 is added to $\frac{1}{2}$ of a number, the result is triple the number.

The answer: 16

Now you show me how to solve this puzzle using an equation.

327. 25% of a 100-gallon tank is filled with a chemical that has to be in water to keep it cool for a nuclear reactor. The scientists have to add more of the chemical to the 100 gallons to get the mixture to 40% of the total container. How much more of the chemical in its 100% form has to be added to the 100 gallons to get a 40% mixture?

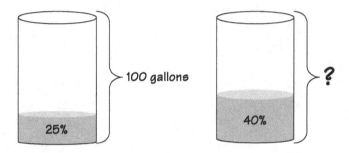

227. 75% of a 100-gallon tank is filled with a chemical that has to be in water to keep it cool for a nuclear reactor. The scientists have to add more of the chemical to the 100 gallons to get the mixture to 40% of the total container. How much more of the chemical in its 100% form has to be added to the 100 gallons to get a 40% mixture?

Answers
Part 1. Numbers and Operations

Whole Numbers

1.

16	9	2	7
6	3	12	13
11	14	5	4
1	8	15	10

Each row, column, and diagonal must total 34, and the four squares in the middle (2 × 2) also must total 34.

2. a. 400. $11^2 = 121$ and $20^2 = 400$

   b. 64. $3^3 = 27$ and $4^3 = 64$. If you reasoned that 27 is 9 times 3 and 9 times 4 is 36, that is correct also.

   c. Heptagon

3. The missing number is 71. If you place these numbers closer together, you see that these are the odd numbers in succession starting with 1.

   1 3 5 7 9 11 13 15 1<u>7</u> <u>1</u> (9)

4. The missing number is 48. Take the difference of the two lower numbers and multiply that difference by the top number to arrive at the number in the middle.

5. a.
$$\begin{array}{r} 7483 \\ + 7455 \\ \hline 14938 \end{array}$$

A = 4   L = 5
B = 7   M = 9
E = 3   S = 8
G = 1

b.
$$\begin{array}{r} 983 \\ 86 \\ + 8 \\ \hline 1077 \end{array}$$

M = 9   B = 1
A = 8   U = 0
D = 3   L = 7
S = 6

6. His average speed will be 6 mph. Hard to believe? Let's say the first half of the trip is 12 miles. 4 mph will take him 3 hours to go halfway. The second half of the trip is his return of 12 miles, and that will take him 1 hour. 4 hours total to cover 24 miles is 6 mph.

7. 31. If there are 32 entrants, 31 of them had to lose in a singles match. So, there are a total of 31 matches.

8. The 500th number is 2,999.

A chart similar to the chart below may have helped you see how to arrive at the solution.

Number	5	11	17	23	29	35	41	...
Position	1	2	3	4	5	6	7	

The pattern is to multiply the position of the number by 6, then subtract 1.

$500 \times 6 = 3000$

$3000 - 1 = 2999$

9. The number is 4. Each of the seven numbers has digits that total 15 when added together.

10. They will be 40 mers apart 1 second before they collide. The two particles approached each other at a speed of 15 + 25, or 40 mers a second, so 1 second before they collide, they will be 40 mers apart.

11. Laughing All the Way to the Bank

12. Eight cans of dog food will feed 40 puppies and 18 dogs. If one can of dog food feeds 8 puppies, then 8 cans would feed 64 puppies. We need to feed 40 pups, leaving 64 – 40, or 24 puppies that have to be "converted" into dogs. Pups are in a 4-to-3 ratio to dogs, so 24 puppies equal 18 dogs.

13.

4	×	3	−	9	+	8	11
+	■	+	■	−	■	×	
8	×	9	+	3	−	4	71
+	■	×	■	+	■	−	
9	×	4	+	8	−	3	41
×	■	−	■	×	■	+	
3	+	8	+	4	×	9	47
39		31		38		38	

14. ☆ = 8

⬠ = 1

☐ = 5

✚ = 6

15. The missing numbers, from left to right, are 20 and 4.

The last number in each row is one less than the first number and the first number times the last number (in each row) is the middle number.

16. The first number is 2. This is really two sequences in one. Start with the third number in the sequence, which is 5 . . . and look at every other number:

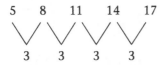

5   8   11   14   17

3   3   3   3

The difference between every other number is 3. Now look at every other number starting with 1.

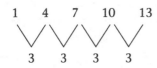

1   4   7   10   13

3   3   3   3

You can see the difference between these numbers is 3 as well.

17. The number 7 goes in the lower right corner. The number 21 goes in the center.

In each box, multiply the top two numbers and put their 2-digit result in the two boxes on the bottom of each box. Then add all four numbers to arrive at the middle number in each box.

18. The missing number is 625. The sequence can be figured as follows:

5^1 5^2 5^3 $\underline{5^4}$ 5^5 5^6

19. The "odd" number out is 13,754. It is the only even number.

20. There are 12 zeros in a thousand thousand million. A thousand thousand is one million, or 10^6. Take that times another million to get $10^6 \times 10^6 = 10^{12}$. The result looks like this: 1,000,000,000,000.

21. Line of Scrimmage

22. There are 5 Crizellas and 3 Frizellas.

$$5 \times 4 = 20 \text{ heads}$$
$$\underline{11 \times 3 = 33 \text{ heads}}$$
$$53 \text{ heads}$$

23. The number 22 in Base$_{10}$ would be the number 24 in Base$_9$. Our Base$_{10}$ system is built on the powers of 10. As an example, let's see how our Base$_{10}$ number of 57 looks to a mathematician.

10^2	10^1	10^0
	5	7

$$5 \times 10^1 = 50$$
$$\underline{7 \times 10^0 = 7}$$
$$57$$

(Any number to the zero power is 1.)

Now, let's see how to convert 22 to Base$_9$.

9^2	9^1	9^0
	2	4

$$9^1 \times 2 = 18$$
$$\underline{9^0 \times 4 = 4}$$
$$22$$

24. 192 total individual digits. Sometimes it helps to make a chart.

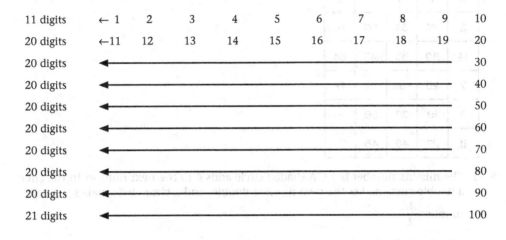

11 digits	← 1	2	3	4	5	6	7	8	9	10
20 digits	←11	12	13	14	15	16	17	18	19	20
20 digits	←									30
20 digits	←									40
20 digits	←									50
20 digits	←									60
20 digits	←									70
20 digits	←									80
20 digits	←									90
21 digits	←									100

1 row with 11 → 11
8 rows with 20 → 160
1 row with 21 → 21
192

25. Starting with the 9 and the 20 on the bottom row—which total 29—the sum of any two consecutive numbers in each row will be found directly above those two numbers in the next row up. The sum appears to "split the difference" of the two numbers beneath it.

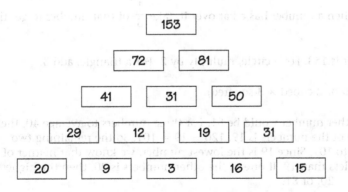

```
              153
          72      81
       41    31    50
     29   12    19    31
   20   9    3    16   15
```

26. b. Larry's second statement is false. His Bingo row was the top row.

27. c. Jane's third statement is false.

B	I	N	G	O
2	22	32	50	61
14	30	35	47	64
7	25	Free	52	72
5	19	37	59	70
11	27	45	48	75

28. a. The missing number is 14. A shaded circle adds 4 to the next number in the circle, a double circle makes the next number double, and a clear circle makes the next number $\frac{1}{2}$.

    b. **31 32** If there is no line under a number, then add 5 to it to get the next number. Put a bar under that new number. Add 1 to any number with a bar under it to get the next number. The new number following a number with a bar will not have a bar under it.

    c. **9** If a number has no bar under it, then add 4 to it to determine the next number. The new number should have a bar under it. When a number has a bar under it, then the next number is $\frac{1}{2}$ the value of that number. This new number should have a bar over it (but not underneath it).

    d. **16** If a number has no bar over it, then to get the following number, double it and put a bar over it. When a number has a bar over it, take $\frac{1}{3}$ of that number to get the following number.

    e. The missing number is 153. For a circle, multiply by 2. For a triangle, add 3.

29. The answer is 72 (8 × 9 or 2-cubed × 3-squared).

30. The highest possible other number would be 81. For three numbers to average 40, their sum must be 120. One of the numbers is 19. 120 − 19 = 101. So the remaining two numbers must add up to 101. Since 19 is the lowest number, we know that neither of the other numbers can be less than 19. If one of the other numbers is 20, then the highest number would be 101 − 20, or 81.

31. V22. The difference between the numbers and the letters is 1, 2, 3, 4, 5, and 6, as shown below.

A	B	C	D	E	F	G	H	I	J	K	L	M	N	O	P	Q	R	S	T	U	V	W	X	Y	Z
1	2	3	4	5	6	7	8	9	10	11	12	13	14	15	16	17	18	19	20	21	22	23	24	25	26

32. Holy Smoke

33. There are 30 numbers (1–30) on Casey's number wheel. In an evenly spaced circle with an even number of objects, if you subtract the opposite numbers in the circle and then multiply by 2, you'll know the total number of objects on the circle. For example, look at a clock.

$$\left.\begin{array}{l} 12 - 6 = 6 \\ 9 - 3 = 6 \\ 7 - 1 = 6 \\ 11 - 5 = 6 \end{array}\right\}$$ Multiply by 2 = 12 numbers on a clock.

34. The number 240 doesn't belong with the rest. All the rest are evenly divisible by 9.

35. The numbers around the triangles are the consecutive prime numbers. Add those three numbers in each figure and multiply by 2 to get the number in the middle.

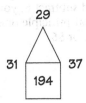

36. 42 students took neither biology nor chemistry.

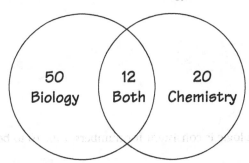

The way to look at this is to consider the 50 + 20 students taking biology or chemistry minus the 12 who are taking both. That gives us 58. You then subtract 58 from 100 to get the answer of 42 students who took neither subject.

37. There are 9 ways.

8 dimes – 0 nickels	3 dimes – 10 nickels
7 dimes – 2 nickels	2 dimes – 12 nickels
6 dimes – 4 nickels	1 dime – 14 nickels
5 dimes – 6 nickels	0 dimes – 16 nickels
4 dimes – 8 nickels	

38. There are 1,000 different possible phone numbers. There are 10 possibilities for the number after "36." With each of those 10 numbers, there are 100 possibilities because they have to be coupled with every number from 00 to 99 after "48." So you multiply 10×100 to get 1,000.

39. c. 222. These are the consecutive even numbers starting with 2, but separated out into 2-digit and 3-digit numbers.

<u>2 4 6 8</u> <u>10 12 14</u> <u>16 18 20</u> <u>22 24</u>

40. a. 2. If two typists can type two pages in 4 minutes, that means one typist can type one page in 4 minutes. So, in 20 minutes one typist can type five pages, and two typists can type ten pages.

41. On the first day, 24 fish were caught. If 200 bass were caught in 5 days, the average is 40 fish a day ($200 \div 5 = 40$). If you start with 40 on the middle day and subtract 8, you get 32 for the second day. Add 8 to get 48 on the fourth day. Now apply that principle one more time to get $32 - 8 = 24$ on the first day. And on the last day, $48 + 8$, or 56.

24	32	<u>40</u>	48	56
1st Day	2nd Day	3rd Day	4th Day	5th Day

42. The impossible scores are 21 and 33.

43. It is possible, and the answers are:

$3{,}125 \times 32 = 100{,}000$

$15{,}625 \times 64 = 1{,}000{,}000$

Take a look at this chart to see how the logic is consistent for numbers from 10 to beyond 1,000,000.

$2^1 \times 5^1 = 2 \times 5 = 10$

$2^2 \times 5^2 = 4 \times 25 = 100$

$2^3 \times 5^3 = 8 \times 125 = 1{,}000$

$2^4 \times 5^4 = 16 \times 625 = 10{,}000$

$2^5 \times 5^5 = 32 \times 3125 = 100{,}000$

$2^6 \times 5^6 = 64 \times 15625 = 1{,}000{,}000$

$2^7 \times 5^7 = ? \times ? = 10{,}000{,}000$

44. A Snail's Pace

45. a. Any power of 10 divided by 9 will always have a remainder of 1. Try some different examples.

   b. 4^{69} divided by 10 has a remainder of 4. Take a look at this chart:

   $4^2 = 16$ – when divided by 10 it has a remainder of 6.

   $4^3 = 64$ – when divided by 10 it has a remainder of 4.

   $4^4 = 256$ – when divided by 10 it has a remainder of 6.

   $4^5 = 1,024$ – when divided by 10 it has a remainder of 4.

   $4^6 = 4,096$ – when divided by 10 it has a remainder of 6.

   $4^7 = 16,384$ – when divided by 10 it has a remainder of 4.

As you can see, any power with a base of 4 has either 6 or 4 as a remainder when divided by 10. Try this with other bases. Below are some "rules of divisibility" which will help you determine if a number is divisible by a certain integer from 1 through 9.

Rules of Divisibility:

   1. *All numbers are divisible by 1.*

   2. *All even numbers are divisible by 2.*

   3. *A number is divisible by 3 if its digits add up to a number that is divisible by 3.*

   4. *A number is divisible by 4 if the last two digits are divisible by 4.*

   5. *A number is divisible by 5 if it ends in 0 or 5.*

   6. *A number is divisible by 6 if it is divisible by 2 and 3.*

   7. *A number is divisible by 7 if you can take the last digit, double it and subtract it from the rest of the number – and that number is divisible by 7.*

   8. *A number is divisible by 8 if the last three numbers are divisible by 8.*

   9. *A number is divisible by 9 if you can divide the sum of the digits by 9.*

   c. 333,333 and 666,666. A number is divisible by 9 if you can divide the sum of the digits by 9 (see rule 9 above).

46. d. 857142. These are the same digits being reshuffled as a result of multiplying 142857 × 1 through 6.

$1 \times 142{,}857 = 142{,}857$

$2 \times 142{,}857 = 285{,}714$

$3 \times 142{,}857 = 428{,}571$

$4 \times 142{,}857 = 571{,}428$

$5 \times 142{,}857 = 714{,}285$

$6 \times 142{,}857 = 857{,}142$

47. You know that four numbers will average 8 if their sum is 32.

   a. 5.

   b. Four different combinations of four numbers will average 8:

   9,9,8,6

   9,9,7,7

   8,8,8,8

   9,9,9,5

   9,8,8,7

48. c. 64,000

   $40^4 = 1{,}600 \times 1{,}600 = 2{,}560{,}000$

   $\dfrac{2{,}560{,}000}{40} = 64{,}000.$ You would have to add 64,000 40's to get 40^4.

49. ① A   18   30   100

        ↓   ↓   ↓

     B   **37**   **61**   **201**

   Row B is 2 times Row A plus 1.

   ② A   10   15   30

        ↓   ↓   ↓

     B **100**   **225**   **900**

   Square each number in Row A to get the respective number in Row B.

   ③ A   22   30   34

        ↓   ↓   ↓

     B   **11**   **15**   **17**

   Starting with the number 2, every other number is even and 4 greater than the preceding even number. Starting with the number 3 in Row A, every other number is odd and 4 greater than the preceding odd number.

50. The five consecutive even numbers are 124, 126, 128, 130, and 132. Divide 640 by 5 to get 128, which is the average of the five numbers and the middle number itself. That means two of the numbers will be greater and two will be smaller to keep the average at 128. So the two consecutive even numbers that are larger are 130 and 132, and the two consecutive even numbers that are smaller are 124 and 126.

51. Here are the six consecutive even numbers:

100   102   104   106   108   110

105

105 is the average of six numbers totaling 630, but it's an odd number, so you use the even numbers before and after 105. 630 was used because there are not six consecutive even whole numbers that total 640!

52. Each box is 6 pounds, and the circle is 10 pounds. The five weights on the right side of the mobile have to total 40 pounds to balance the two triangles on the left side. Since the hourglass is 14 pounds, the box above it must be 6 pounds so they total 20 pounds, which will balance the 20 pounds where the circle, box, and star are. We know the star is 4 pounds and the box is 6 pounds, so the circle must be 10 pounds to balance the smaller mobiles contained within the big mobile.

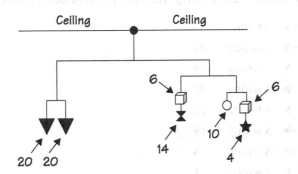

53. A Pazooto costs 20¢.

A Fazooto costs 30¢.

A Razooto costs 40¢.

If a Fazooto and a Pazooto together cost 50 cents and a Pazooto and a Razooto together cost 60 cents, a Razooto is 10 cents more than a Fazooto. Since a Razooto and Fazooto cost 70 cents together, a Razooto must be 40 cents and a Fazooto is 30 cents. That means a Pazooto must be 20 cents (50 – 30 and 60 – 40 both equal 20).

54. A and B are 8 and 4, or 4 and 8. The value of C is 9 because 14 – 5 = 9. That means the intersection of *just* circles A and B must be 12. 21 – 9 (the value of C) = 12. The only two numbers A and B can be are 4 and 8 (or 8 and 4) because D = 5, so you can't use 5 and 7. You can't use 6 and 6 because they can't both be the same number. And you can't use 9 and 3 because C = 9.

55. Take Into Account

56. The missing number is 21. Add the three numbers on the outside of each triangle and then reverse the digits in the sum.

57. The missing number is 9. Multiply the first number in the row by the second number in the row, then add the digits in the sum to get the number in the third column.

$7 \times 4 = 28 \rightarrow 2 + 8 = 10$

$3 \times 6 = 18 \rightarrow 1 + 8 = 9$

$2 \times 5 = 10 \rightarrow 1 + 0 = 1$

$6 \times 8 = 48 \rightarrow 4 + 8 = 12$

$8 \times 9 = 72 \rightarrow 7 + 2 = \underline{\quad 9 \quad}$

58. You could take advantage of this opportunity if you determine that each consonant in each of the items' names is worth $40 and, therefore, a pair of shoes would sell for $120.

59. $36 \times 196 = 7056$. A number times itself is equal to 7056. You can quickly see that it has to be larger than 80×80 (6400) and has to end in a 4 or 6 because the last digit of 7056 is 6. Trying 84×84 will give the correct result: $84 \times 84 = 7056$.

60. The answer is 7 correct and 3 wrong. Here is a list of possible scores:

10 correct	→ 0 wrong	50
9 correct	→ 1 wrong	43
8 correct	→ 2 wrong	36
7 correct	→ 3 wrong	29
6 correct	→ 4 wrong	22
5 correct	→ 5 wrong	15
4 correct	→ 6 wrong	8
3 correct	→ 7 wrong	1
2 correct	→ 8 wrong	–6
1 correct	→ 9 wrong	–13
	10 wrong	–20

At every level there is a difference of 7. A quick way to find out how many are right and wrong is to subtract the score from 50 and divide by 7. The result will give you the number of wrong answers. You then subtract that number from 10 to get the number of correct answers.

61. $x = 10$. Look at the row beginning with 14. $14 + 8 + 11 = 33$ and you still need a number. So each row, column, and diagonal will have a total greater than 33. Now look at the column beginning with 3. $3 + 8 + 13$ is equal to 24. Therefore, x has to be greater than 9 (because $9 + 24 = 33$). You've already used 11, 12, 13, 14, 15, and 16. The next highest number is 10. So $x = 10$ and each row, column, and diagonal totals 34.

9	6	3	16
4	15	10	5
14	1	8	11
7	12	13	2

62. Column H. Since each row is a multiple of 8 and 8 goes into 1,000 (125 times) with no remainders, it will go under H. If the number were 1,001, it would go under A because $\dfrac{1,001}{8}$ has a remainder of 1. If the number were 1,004, the number would go under D because $\dfrac{1,004}{8}$ has a remainder of 4.

63. c. 51. The first three digits add up to 12, the second group of two digits adds up to 10, and the third group of two digits adds up to 6 in all six rows. Any two-digit number that adds up to 6 would work, but 51 is the only one of the choices for which this is true.

64. It took 11 years. Then it doubled its length once more to reach its maximum in the 12th year.

65. He is 8 years old. Two years from now he will be 10. $2 \times 10 = 20$. Two years ago he was 6.

$6 \times 2 = 12.$ $20 - 12 = \underline{8}$.

66. When Push Comes to Shove

67. You would write it 120 times. Remember—every number you write begins with 4, so that's 100. Ten of the numbers have another 4 in the tens place (440, 441, etc.), and ten numbers have a 4 in the ones place (404, 414, etc.).

68. 27. Every number after __27__ can be formed by some combination of 5's and 8's.

69. 1, 5, 10, 10, 5, 1. The sum of the numbers in this row is 32. The middle numbers are both 10s. You can view this in several ways. One way is to notice how the numbers progress from top to bottom vertically (catercorner or diagonally). For example: Look at row 1 and follow the numbers down right to left. You will see that they are in sequential order (1, 2, 3, 4, and so on). Look for different patterns in the different diagonals.

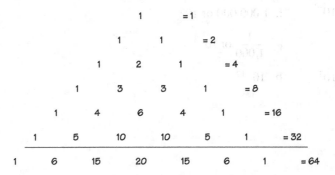

70. d. 756. It is the only number divisible by 4. All the numbers in the sequence are divisible by 4.

71. The missing result is 47. Multiply the two numbers in each row, and then subtract their sum. $63 - 16 = 47$.

72. 16,000 miles. If Brad had only two tires, then each tire would have 24,000 miles of use for a total of 48,000 miles. Since he has three tires and they received the same usage, they each were used for 16,000 miles ($16,000 \times 3 = 48,000$ miles).

73. The missing number is 46. Add the four numbers in the squares surrounding each diamond—and double the sum to get the answer in the middle.

74. There are 301 numbers. Take a look, using a chart.

Position	1	2	3	4	5	6	7	8	9	10	11
Number	0	3	6	9	12	15	18	21	24	27	30

Notice that if you divide the numbers by 3 and add 1, you have the position of the number: $900 \div 3 = 300$. Add 1 to get 301.

75.

0	−7	−5
−9	−4	1
−3	−1	−8

Rational Numbers

76. If 72% are business majors, then 28% are not.

If 58% are math majors, then 42% are not.

So 28 + 42, or 70% of the students, *are not* majoring in both.

So 30% or 30 students *are* majoring in both.

77. a. $\frac{1}{10}$ or 10^{-1} e. 1,000 or 10^3

    b. $\frac{1}{100}$ or 10^{-2} f. 1,000,000 or 10^6

    c. 10^9 g. $\frac{1}{1,000}$ or 10^{-3}

    d. 100 or 10^2 h. 10^{-12}

Here is a more complete list of prefixes and their values.

Number	Prefix	Symbol	Number	Prefix	Symbol
10^1	deka-	da	10^{-1}	deci-	d
10^2	hector-	h	10^{-2}	centi-	c
10^3	kilo-	k	10^{-3}	milli-	m
10^6	mega-	M	10^{-6}	micro-	μ
10^9	giga-	G	10^{-9}	nano-	n
10^{12}	tera-	T	10^{-12}	pico-	p
10^{15}	peta-	P	10^{-15}	femto-	f
10^{18}	exa-	E	10^{-18}	atto-	a
10^{21}	zeta-	Z	10^{-21}	zepto-	z
10^{24}	yotta-	Y	10^{-24}	yocto-	y

78. Careful Calculations

79. Here's one way: $\dfrac{4}{4} + \dfrac{4}{4} = 2$

    Did you find another?

80. Her daughter reasoned the puzzle like this: The gas in the tank can be divided into 7 parts. They had used only 5 of the 7 parts and had driven 360 miles. Therefore, for each $\dfrac{1}{7}$ portion, they had driven $360 \div 5$, or 72 miles. Multiply 72 by 7 and you have the answer if the tank had been completely full. $72 \times 7 = 504$ miles.

81. Here are two possibilities, and there may be more:

    a. $50 + 49 + \dfrac{1}{2} + \dfrac{38}{76} = 100$

    b. $97 + \dfrac{8}{12} + \dfrac{4}{6} + \dfrac{5}{3} = 100$

82. Here's how to make 42 using just 4s:

    $$\dfrac{4 \times 4}{.4} + \sqrt{4}$$

    Can you find other ways to come up with 42?

83. c. They are equal.

84. c. 11. You can view the puzzle by breaking it down.

    $\dfrac{10^{11}}{10^{10}} = 10 \qquad \dfrac{10^{10}}{10^{10}} = 1 \qquad \dfrac{10^{11}}{10^{10}} + \dfrac{10^{10}}{10^{10}} = 10 + 1 = 11$

85. a.

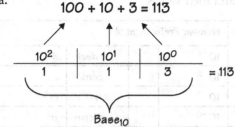

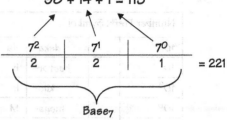

b.

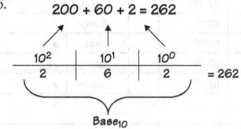

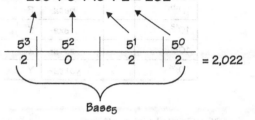

c.

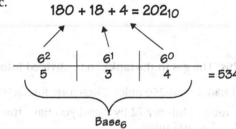

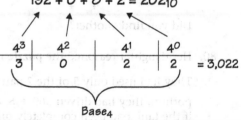

86. a. $\frac{5}{6}$. 270° is $\frac{3}{4}$ of a circle (360°). $\frac{5}{6}$ of 360° is 300°.

b. Octagon. A hexagon has 6 sides; a triangle has half that number—3 sides. A square has 4 sides; twice that is 8—the number of sides of an octagon.

c. 625. $3^2 = 9$ and $9^2 = 81$

$5^2 = 25$ and $25^2 = 625$

87. One Ponplo equals .23339 Konklos. Divide 1 by 4.28466.

$$\frac{1}{4.28466} = .23339$$

88. It's on the Tip of My Tongue

89. e. $\frac{1}{10} \div .1$ is the same as $\frac{1}{10} \times 10$ or 1. All the other values are fractions.

90. A decimal point. 6.7

91. 60%. Almors increased by 20% becomes an increase of 120%, or 1.2, which is equal to Brons. 1.2 or 120% decreased by 50%, or $\frac{1}{2}$ is equal to .6, which is Choops. So Choops is 60% of Almors.

92. $1.20. If 7 pencils cost 30¢ more than 5 pencils, then 2 pencils would cost 30¢ and 1 pencil would be 15¢. $15 \times 8 = \$1.20$.

93. The letter D.

    A   B   C   D   E   F   G   H   I   J

    There are 10 letters counting A and J. $\frac{2}{5}$ of 10 is 4. The fourth letter is D.

94. .0016 or $\frac{16}{10,000}$ or $\frac{1}{65}$

    $.01 \times .02 \times 12 \times \frac{2}{3} = .0016$

95. Here are two answers:

    $$\begin{array}{r} 93.2 \\ -41.0 \\ \hline 52.2 \end{array} \qquad \begin{array}{r} 92.3 \\ -40.1 \\ \hline 52.2 \end{array}$$

96. e. $\dfrac{27}{.004}$

97. 180. 3 times 60 is 300 percent more.

98. $\frac{2}{7}$ to $\frac{5}{7}$ or 2 to 5

99. Caught Off Balance

100. 75% of Mike's pitches were strikes. He would have had to throw 38 strikes to be at 85%.

    $$\frac{33}{44} = \frac{x}{100\%} \qquad\qquad \frac{x}{44} = \frac{85\%}{100\%}$$

    $$\frac{44x}{44} = \frac{33 \times 100}{44} \qquad\qquad \frac{100x}{100} = \frac{44 \times 85}{100}$$

    $x = 75\%$ $x = 37.4 \to 38$ (since you can't throw a partial pitch, this is rounded up)

101. a. Close to $\frac{1}{2}$.

    When dividing fractions, invert and multiply $\left(\frac{1}{3} \times \frac{11}{13} \times \frac{7}{4}\right)$.

    $\dfrac{77}{156} = .4935897$

102. $\frac{3}{4}$ and $\frac{5}{6}$

    $\left.\begin{array}{l} \dfrac{3}{4} = .75 \\[2mm] \dfrac{5}{6} = .833 \end{array}\right\}$ .8 is between these two fractions.

103. a. Work from the inside out: $\sqrt{256} = 16$; $\sqrt{16} = 4$; $\sqrt{4} = \pm 2$

   b. This is the same as $x^{\frac{1}{8}}$.

104. $\frac{1}{13}$. If you look closely, this is really two sequence puzzles. Look at the first fraction, $\frac{1}{3}$ . . . and then every other number. You'll see that it is the successive odd counting numbers in the denominators.

$$\frac{1}{3} \quad \frac{1}{5} \quad \frac{1}{7} \quad \frac{1}{9} \quad \frac{1}{11} \quad \frac{1}{13}$$

The other sequence starts with $\frac{1}{2}$ and proceeds: $\frac{1}{2} \quad \frac{1}{4} \quad \frac{1}{8} \quad \frac{1}{16} \quad \frac{1}{32}$

105.

Ship To:	Mary Ellsbeth		Date:	January 3, 2010
	16 Forest Drive			
	Townsville, Alaska 02671			

QUAN.	DESCRIPTION	UNIT PRICE	TOTAL PRICE
2	Brite-Lite Gadget	$ 16.25	$ 32.50
1	Useful Gidget	$ 7.30	$ 7.30
2	Gadget Attack! Game	$ 12.95	$ 25.90
3	Gidget Packs	$ 4.85	$ 14.55
		Sub-Total	$ 80.25
		Shipping:	$ 8.00
		Total Order:	$ 88.25
		Total Due:	Paid in Full

106. c. $13\frac{1}{8}$

   If 7 numbers have an average of 6, then the sum of all 7 numbers $= 42 \rightarrow (7 \times 6)$.

   If 6 numbers have an average of 7, then the sum of all 6 must be $42 \rightarrow (6 \times 7)$.

   If 3 numbers have an average of 42, then the sum of all 3 numbers must be $126 \rightarrow (3 \times 42)$.

$$\begin{array}{r} 42 \\ 42 \\ +126 \\ \hline 210 \end{array} \qquad 16\overline{)210} = 13\frac{1}{8}$$

107. $\frac{5}{17}$. Here's how to work it out:

$$30\% = .3$$

$$.3 \times \frac{25}{17} = \frac{7.5}{17}$$

$$\frac{7.5}{17} \div \frac{3}{4} = \frac{7.5}{17} \times \frac{4}{3} = \frac{30}{51}$$

$$\frac{30}{51} \times \frac{1}{2} = \frac{30}{102} = \frac{15}{51} = \frac{5}{17}$$

108. $\frac{1}{3} \times \frac{1}{3} = \frac{1}{9}$

$\frac{1}{2} \times \frac{1}{2} = \frac{1}{4}$

$\frac{1}{5} \times \frac{1}{5} = \frac{1}{25}$

$\frac{1}{9} \times \frac{1}{4} = \frac{1}{25} = \frac{100}{900} + \frac{225}{900} + \frac{36}{900} = \frac{361}{900}$

109. Friday. Here's how a chart can help you solve this puzzle:

Tues.	Wed.	Thurs.	Fri.	Sat.	Sun.	Mon.
1	2	3	4	5	6	7
8	9	10	11	12	13	14
15	16	17	18	19	20	21
22	23	24	25	26	27	28
29	30	31	32	33	34	35

Notice that if you divide 7 into any of the numbers, the remainder (except Monday, which has no remainder) is the *position* of the day you are seeking. Example: $25 \div 7 = 3\frac{4}{7}$. The fourth day from Tuesday is *Friday*. Example: $31 \div 7 = 4\frac{3}{7}$. The third day from Tuesday is *Thursday*. $200 \div 7 = 28\frac{4}{7}$ so 4 days from Tuesday is *Friday*, which is the correct answer.

110. Fall Over Backwards

111. d. 1.2. Let z = zeeko and t = teeko: $\frac{1}{2}z = .6t$

$z = 1.2t$ $\frac{z}{t} = \frac{1.2t}{t} = \frac{6}{5}$

112. b. $\frac{19}{211}$

113. 55%. Let's say they play 90 total games. The first third of their games would be 30 games. If they had won 40% of those games, they would have $30 \times .4$, or 12 wins. They need 45 total wins to finish with an even record, so they need to win 33 more games. They have 60 games left: $\frac{33}{60} = .55$. They have to win 55% of their remaining games to finish with 45 wins.

114. 72%. A increased by 30% is 1.3.

$$\begin{array}{r} 1.30 \\ -.52 \\ \hline .78 \end{array}$$

1.3 decreased by 40% is $1.3 \times .4 =$

$$.78 \times .20 = \begin{array}{r} .78 \\ +.156 \\ \hline .936 \end{array}$$

$$.936 = D$$

$$\frac{.936}{1.3} = 72\%$$

115. a. $\underline{5^{10}}$ 5^{10} is 9,765,625 10^5 is 100,000

   b. They are the same.

   c. They are the same.

116. b. $\dfrac{24}{101}$

$$\frac{1}{4+\dfrac{1}{4}} = \frac{1}{\dfrac{5}{4}} = \frac{1}{1.25} = .8. \text{ Then move to the next step} \ldots$$

$$\frac{1}{4+(.8)} = \frac{5}{24}, \text{ then}$$

$$\frac{1}{4+\dfrac{5}{24}} = \frac{1}{\dfrac{96}{24}+\dfrac{5}{24}} = \frac{1}{\dfrac{101}{24}} = \frac{24}{101}$$

117. Partner A gets $15,000.

Partner B gets $20,000.

Partner C gets $25,000.

The total amount invested was $12,000.

Partner A invested $3,000, which is $\dfrac{1}{4}$.

Partner B invested $4,000, which is $\dfrac{1}{3}$.

Partner C invested $5,000, which is $\dfrac{5}{12}$.

$\dfrac{1}{4} \times 60,000 = 15,000$ for partner A

$\dfrac{1}{3} \times 60,000 = 20,000$ for partner B

$\dfrac{5}{12} \times 60,000 = 25,000$ for partner C

118. $\frac{1}{19}$ should be $\frac{1}{18}$. The denominators are all factors of 36.

119. Here's one way. Did you find another?

Step 1. Press 125%

Step 2. Press 200% four times

Step 3. Press 5%

120. $\frac{3543}{1550}$

$$\frac{1}{4\frac{3}{7}} + \frac{1}{3\frac{11}{13}} + \frac{1}{\frac{5}{9}} =$$

$$\frac{1}{\frac{31}{7}} + \frac{1}{\frac{50}{13}} + \frac{1}{\frac{5}{9}} =$$

$$\frac{7}{31} + \frac{13}{50} + \frac{9}{5} =$$

$$\frac{350}{1550} + \frac{403}{1550} + \frac{2790}{1550} = \frac{3543}{1550}$$

121. Fishing for Compliments

122. The new mixture is 10 units of 36 units lemonade and 26 units of 36 units iced tea. Let's take a different perspective on this puzzle. Since the second container is twice the size, make it into two containers—each one-half the size of the larger container.

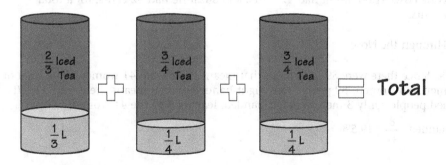

Notice that if each container was 12 units large, the numbers are easy to work with.

The 1st container would have 8 units of iced tea and 4 units of lemonade.

The 2nd container would have 9 units of iced tea and 3 units of lemonade.

The 3rd container would have 9 units of iced tea and 3 units of lemonade.

So, in 36 total units, you have 26 units of iced tea and 10 units of lemonade.

$$\frac{26}{36} = \frac{13}{18} = \text{Iced Tea}$$

$$\frac{10}{36} = \frac{5}{18} = \text{Lemonade}$$

123. The chance would be 19.36%. To get the answer, multiply the 20% of people who didn't like chocolate times the 88% of the people who are older than 15 to get .20 × .88 = .1936, or 19.36%.

124. 4. 50% × 200% is 100%; 100% of 100% is 1; $1 \div 25\% = 1 \div \frac{1}{4} = \underline{4}$.

125. 8, 13, 26, 52. The next number would be 104. You can approach this in several ways. One interesting way is to find the number *closest* to 109 that has a remainder of 5, then work backward using the multiples of that number. The number 104 is the closest to 109 that will leave a remainder of 5. Then take $\frac{1}{2}$ of 104 to get 52, $\frac{1}{2}$ of 52 to get 26, $\frac{1}{2}$ of 26 to get 13, and 5 less than 13 is 8.

126. Here's another way.

$$
\begin{aligned}
4 \text{ quarters} &= \$1.00 \\
4 \text{ dimes} &= \$\ .40 \\
\underline{12 \text{ nickels}} &= \underline{\$\ .60} \\
20 \text{ coins} &= \$2.00
\end{aligned}
$$

127. The cone costs $1.00. Mimi had 47 cents, and Suzanne had 52 cents, for a total of 99 cents.

128. Pay Through the Nose

129. 19.5%. Since there were 59 men total, that means there were 41 women, for a total of 100 men and women. 89 people were right-handed, which means there were 11 left-handed people. Only 3 men were left-handed, leaving 8 of the 41 women to be left-handed: $\frac{8}{41} = 19.5\%$

130. $\frac{5}{8}$

$$\frac{A}{B} = \frac{3}{4} \quad \text{and} \quad \frac{B}{C} = \frac{5}{6}$$

$$4A = 3B \qquad 5C = 6B$$

$$A = \frac{3}{4}B \qquad C = \frac{6B}{5}$$

$$\frac{A}{C} = \frac{\frac{3}{4}B}{\frac{6}{5}B} = \frac{3}{4} \times \frac{5}{6} = \frac{15}{24} = \frac{5}{8}$$

131. a. $\dfrac{64}{4} = \underline{16}$

    b. $\dfrac{1}{2} \times \dfrac{1}{3} \times \dfrac{1}{5} \times \dfrac{2}{5} \times \dfrac{8}{10} = \dfrac{16}{1500} = \dfrac{8}{750} = \dfrac{4}{\underline{375}}$

    c. $120 + 2400 \times \dfrac{1}{2} = 120 + 1200 = \dfrac{1320}{\frac{1}{4}} = 1320 \times 4 = \underline{5280}$

132. 5,000 cubic feet of water. When the pool is full in the morning, it has 100 ft. × 50 ft. × 6 ft., or 30,000 cubic feet of water. But it loses $\dfrac{1}{6}$ of its water per day, from a depth of 6 ft. to 5 ft. So, $\dfrac{1}{6}$ of 30,000 is 5,000 cubic feet.

133. $\dfrac{1}{4}$ or 1 to 4. The class has $\dfrac{4}{5}$ female and $\dfrac{1}{5}$ male students, so the ratio is $\dfrac{1}{5}$ to $\dfrac{4}{5}$ or $\dfrac{1}{4}$ (1 to 4).

134. 19%. Take a look at the first 100 numbers. 4 appears 20 times (if you count both 4's in 44), but the question asked for integers containing at least one 4 so we don't count both 4's in 44. That leaves 19 fours in the first 100 numbers. That ratio exists to infinity.

135. 8 minutes. In one minute, $\dfrac{3}{4}$ or $\dfrac{6}{8}$ of the sink is filled while $\dfrac{5}{8}$ of the sink is emptied. That means that $\dfrac{1}{8}$ of the sink is filled in one minute. So it takes 8 minutes to fill the sink.

136. Here's one way: He laid down the rope against the side of the shed that was 7 ft. long and marked the result. Then he took the 7 ft. section and laid it down against the 5 ft. side of the shed and marked that as well. He now had a 7 ft. piece of rope marked off into two pieces of 5 ft. and 2 ft. Then he laid down the 2 ft. section on the ground three times to mark off the 6 ft. he needed.

137. 28.6%. Jane purchased the jeans for $32 → $40 × 20% = $8 discount off $40 (40 − 8 = 32). She purchased the shirt for $18 → $30 × 40% = $12 discount off $30 (30 − 12 = 18). She would have had to pay $70 full price for both items but paid $50 total ($32 + $18 = $50).

    $\begin{array}{r} 70 \\ -50 \\ \hline 20 \end{array}$ $\dfrac{20}{70} = \dfrac{2}{7} = 28.6\%$ total discount

138. Batter's Box

Part II. Geometry and Measurement

Geometry

139. Choice #1 is correct.

140. E will not fold into a cube.

141. B and D can be folded into the cube shown.

142. The hose in illustration B will fill a swimming pool faster. The way to determine this is to compare the area of the opening where the water will come out, which is a circle. The area of a circle is found by multiplying pi (π) by the square of the radius of each circle. ($A = \pi r^2$).

    In illustration A, the area is equal to $16^2\pi + 16^2\pi = 256\pi + 256\pi = 512\pi$ sq. inch.

    In illustration B, the area is equal to $30^2\pi$ or 900π sq. inch. The hose in illustration B will fill a pool much faster.

143. c. *Hexagon* is the odd figure out. The other four figures have four sides. The hexagon has six sides. If you said quadrilateral, that is acceptable, too. Why? Because the other illustrations all have at least one set of parallel lines. A quadrilateral doesn't have to have parallel sides.

144. Line AG is the radius of the circle, so its length is 7 ft. + 5 ft., or 12 ft. If you were to draw line FG, it also would be 12 ft. because it is a radius as well. That means line EH is 12 ft. because it is a diagonal of the rectangle EFGH and equal to line FG.

145. 12 inches. The perimeter of $\triangle$RST is 8 + 11 + 17, which is 36. $\triangle$ABC has a perimeter of 36, since the two triangles are equal. All of the three sides of $\triangle$ABC are equal because an equilateral triangle has equal sides, so 36 ÷ 3 = 12 for each side.

146. Trapezoid. The trapezoid is the only quadrilateral that isn't a parallelogram—meaning that only one pair of sides is parallel and one side is not parallel.

147. A and B

148. b. Parallel to Saymore St.

149. Parking Space

150. There are two: $\square$ ABFE and $\square$ BCDE

151. 165°. Since Z is a 90° angle, that leaves a total of 90° in the other two angles. X + Y = 90°.

We know from the question that $X = \dfrac{Y}{5}$.

So, $\dfrac{Y}{5} + Y = 90°$.

Now, multiply each term by 5: Y + 5Y = 450

6Y = 450

So, Y = 75° and X = 15°. Now, since we know that a straight line is 180°, subtract X from 180°. So 180 − X (15°) = 165°.

152. B is longer. It has 14 different point-to-point segments, as does A, but it has one more diagonal segment compared to A.

A diagonal segment is longer than a horizontal/vertical segment.

153. All are true except the last statement.

154. You might pick #4 because it is the only illustration that does not have at least two equal sides. Or you could pick #1 because it has sides longer than any of the other illustrations. Or you could pick #5 because it has the shortest sides. Can you find other reasons?

155. The 18″ × 18″ pizza from Louie's is the better deal. That pizza has an area of 18″ × 18″ or 324 sq. inches for $15.00. Dominick's Pizza has an area of $(10)^2 \times \pi$ or 314 sq. inches for $16.00. The area of a circle is found by the formula $A = \pi r^2$. The radius of Dominick's pizza is 10″ (half of the 20″ diameter). Pi is 3.14.

156. The perimeter is 47 inches. An isosceles triangle has two equal sides. It can't be two sides with 7 inches and one side with 20 inches or the sides wouldn't meet. So it must be 20 + 20 + 7 = 47 inches.

157. b. XP = XY − PY

X •————— P ————• Y

158.

Polyhedron	FACES	EDGES	VERTICES
Tetrahedron	4	6	4
Octahedron	8	12	6
Dodecahedron	12	30	20
Icosahedron	20	30	12

159. K is the letter with horizontal symmetry.

160. b.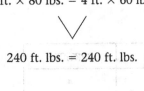

161. c. 960° – each hour is 360°. So 2 hours would be 360° × 2 = 720°. Plus 40 minutes is $\frac{2}{3}$ of 1 hour, or 240°. $\frac{2}{3}$ × 360° = 240°. 720 + 240 = 960.

162. Repeat After Me . . .

Measurement

163. 5:20. The times increase as follows:

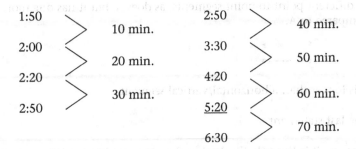

164. You would need 80 square yards.

$$40 \times 18 = \frac{720 \text{ ft.}^2}{9} = 80 \text{ yd.}^2$$

165. c. Greater than 4 units. The distance is exactly 5 units.

166. b. 20 times. 3 × 40 = 6 × 20. The system has to be in balance or equilibrium for the two pulleys to work together.

167. The 80 lb. weight has to be 3 feet from the fulcrum.

3 ft. × 80 lbs. = 4 ft. × 60 lbs.

240 ft. lbs. = 240 ft. lbs.

168. Alicia, soccer practice, Thursday, 4:00 PM

Cindy, piano lessons, Friday, 3:00 PM

Harry, baseball practice, Tuesday, 3:30 PM

Larry, dance lessons, Wednesday, 4:30 PM

Cindy went to her activity on Friday (Clue b). Alicia went to her activity on Thursday at 4:00 PM (Clue d). Since Larry didn't attend baseball practice on Tuesday (Clue c), by elimination, Harry attends baseball practice on Tuesday and Larry attends his activity on Wednesday. Alicia has soccer practice and Cindy's activity starts at 3:00 PM (Clue a). Larry takes dance lessons (Clue f), which starts at 4:30 PM (Clue e). So, by elimination, Harry goes to his activity at 3:30 PM and Cindy takes piano lessons.

169. 12:20.

170. b. 7:50.

171. The 23rd.

172. It will be 9:00. This is easier to see if you make a chart.

Actual Time	Time Read on Clock (12 min. fast/hour)
1:00	1:12
2:00	2:24
3:00	3:36
4:00	4:48
5:00	6:00
6:00	7:12
7:00	8:24
8:00	9:36
9:00	10:48
10:00	12:00
11:00	12:12
12:00	1:24

173. 90 chimes.

Time	Chimes	
	Hour	**Half Hour**
12:00 – 12:30	12	1
1:00 – 1:30	1	1
2:00 – 2:30	2	1
3:00 – 3:30	3	1
4:00 – 4:30	4	1
5:00 – 5:30	5	1
6:00 – 6:30	6	1
7:00 – 7:30	7	1
8:00 – 8:30	8	1
9:00 – 9:30	9	1
10:00 – 10:30	10	1
11:00 – 11:30	11	1
	78 +	12 → 90

174. O-Zone

175. 1.157 days.

$$\frac{100{,}000 \text{ sec}}{60 \text{ sec}} = 1{,}666.7 \text{ minutes}$$

$$\frac{16{,}666.7}{60} = 27.778 \text{ hours}$$

$$\frac{27.778}{24} = 1.157 \text{ days}$$

176. 108 pieces of chocolate can be cut from the sheet. $18'' \times 18'' = 324$ sq. in. Each piece is $2.0'' \times 1.5''$ or 3 sq. in.

$$\frac{324 \text{ sq. in.}}{3 \text{ sq. in.}} = 108 \text{ pieces}$$

177. a. Millennium

b. Micro

c. 10

d. $\frac{1}{9}$

e. Pentagon

f. 3. P. B is the second letter of the alphabet and H is the eighth. D is the fourth letter of the alphabet, and P is the 16th.

$2 : 8 :: 4 : 16$

178. March 21 is the Vernal equinox—exactly 6 months from September 21, and Autumnal equinox.

179. Something that is 1,000 yards square is a square that is 1,000 yards on each side. Something that has an area of 1,000 square yards can be shown by a figure that is 100 × 10, 50 × 20, 40 × 25, etc.

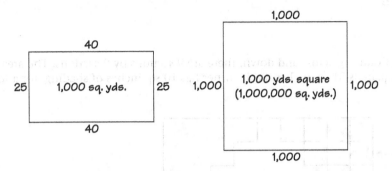

An example of perimeter comparisons might be the figures above. The figure on the left is 40 + 40 + 25 + 25 = 130 yards. The figure on the right is 1,000 + 1,000 + 1,000 + 1,000 = 4,000 yards.

180. They can mow 1,600 square feet in 9 minutes. Sometimes these types of problems are made easier by a picture.

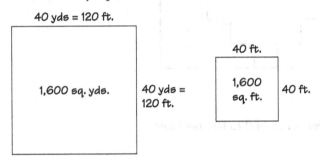

A lawn that is 1,600 square yards is the same as 14,400 square feet.

A lawn that is 14,400 square feet is 9 times the size of a lawn that is 1,600 square feet. So Billy and Barbara can mow the smaller field in $\frac{1}{9}$ the time, or $\frac{81}{9} = 9$ minutes.

181. 200 sq. in. The area of square ABCD is 100 square inches (10″ × 10″). The triangle BCD is $\frac{1}{2}$ that area or 50 square inches. The diagonals of square BDFE form four similar triangles, all of which are 50 square inches. 50 × 4 = 200 sq. inches.

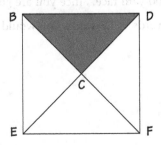

182. The boys will be gone 4 hours and 30 minutes. It takes 45 minutes to walk to the

movie and 30 minutes for a hotdog, so the boys will have to leave Charlie's home at 12:15 PM to make the movie. The movie is over at 3:45 PM. It will take 45 minutes to get home, plus 15 minutes buying milk. So the boys are gone from 12:15 to 4:45, or 4 hours and 30 minutes.

183. c. 720 days.

184. 40 sq. inches. Counting across and down, there are 9 sections by 9 sections. The area of the entire square is 81 sq. inches. Each corner has 10 sq. inches of shading, for a total of 40 sq. inches.

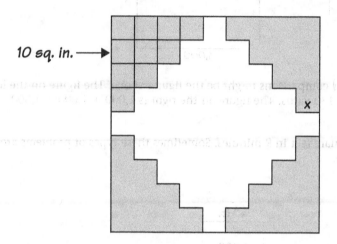

185. Shape Up or Ship Out

186. 741.8 mph. First convert feet per second to feet per hour:

$$1,088\,^{ft}\!/_{sec} \times \frac{60\ sec}{1\ min} = 65,280\,^{ft}\!/_{min}$$

$$65,280\,^{ft}\!/_{min} \times \frac{60\ min}{hr} = 3,916,800\,^{ft}\!/_{hr}$$

Since there are 5,280 feet in a mile, divide the feet by 5,280 to get:

$$3,916,800\ ft \div \frac{5,280\ ft}{hr} = 741.8\ m.p.h.$$

187. 3.5 yards. The person you originally beat would beat the opponent by 12 yards in a 200-yard race (6 yards in a 100-yard race). But you had beaten them by 8 yards in a 200-yard race. So you would beat that person by 14 yards in a 200-yard race. Since you are going to race only $\frac{1}{4}$ that distance at 50 yards, then $\frac{1}{4} \times 14$ yards = 3.5 yards. You would win by 3.5 yards.

Part III. Mathematical Reasoning

Visual

188.

O	O		X
X	X		
		O	
X		O	X

Starting with the second grid, rotate the entire grid 90° clockwise to find the next grid.

189. Shape C is the odd one out.

190. There are 39 blocks.

191. Figure C is the odd one out. The other four figures contain a capital letter F. In C the F is backward (Ⅎ).

192. There are 17 total squares.

$8 \rightarrow 1 \times 1$

$5 \rightarrow 2 \times 2$

$3 \rightarrow 3 \times 3$

$0 \rightarrow 4 \times 4$

$1 \rightarrow 5 \times 5$

193. There are 39 total cubes—20 in one pile, 19 in the other.

194. ▢ ▢ ▢ = MOH   SOCSEH =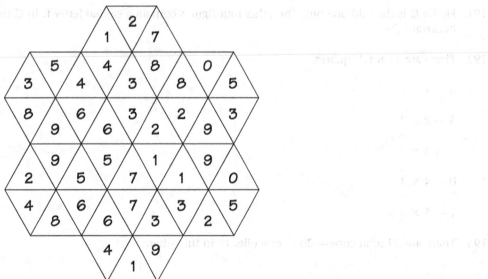

   H = 3   S = Linked

   M = Unlinked   E = Circles

   O = Squares   C = 2

195. THINK

   THANK

   SHANK

   SHARK

   STARK

   START

   SMART

196. Cost of Living Raise (OR Cost of Living Is Rising)

197.

198.

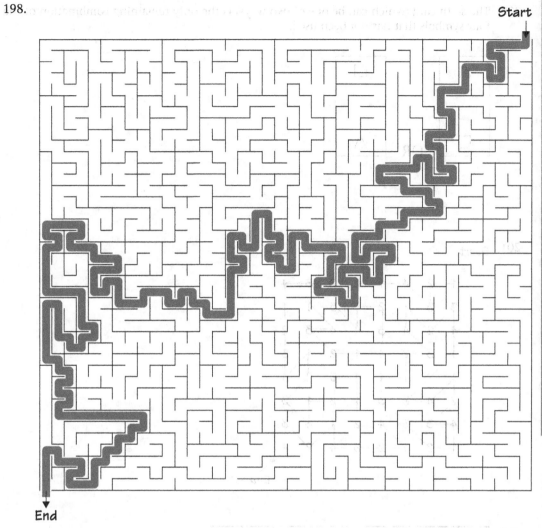

End

199.

21	3	10	12	19
15	17	24	1	8
4	6	13	20	22
18	25	2	9	11
7	14	16	23	5

200. The sixth card (which can be one of two ways) is the only remaining combination of the four symbols that has not been used.

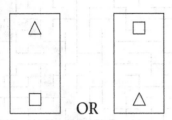

201.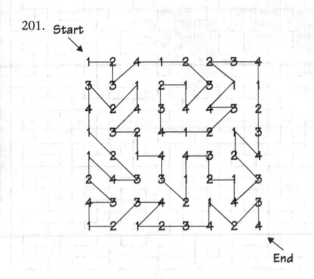

202.

3	9	7	5	3	7	3	6	7
7	2	9	8	6	2	5	8	3
5	8	5	2	3	6	3	7	9
3	6	7	4	7	4	2	6	3
7	4	3	9	5	3	5	2	7
9	2	9	6	8	6	4	6	8
3	6	7	3	5	2	7	5	9
5	4	2	8	3	8	3	6	7
7	3	5	6	9	5	7	2	9

203. The result will look like image c. If you were to cut off the other two corners, instead of those shown, the result would look like d.

204.

108894	A = 6
−56447	E = 9
52447	H = 5
	O = 0
	P = 1
	R = 4
	T = 8
	U = 2

205.

1		0			1		•	2	
•			2	•		1			•
•	•	•		1	1		1	2	
2		3		1		1	•	2	
2		2	•			3	3		•
•	•				2	•	•	•	2
3	•				3			3	
3	4	•	3	•		2	•	3	•
•	•		3	•			•	3	
3	•		1		1			1	

206. You Are a Cut Above the Rest

207.

4	2	3	1	6	9	8	5	7
8	9	5	3	2	7	1	6	4
1	7	6	5	8	4	9	2	3
9	4	8	6	7	3	5	1	2
2	5	1	9	4	8	7	3	6
3	6	7	2	1	5	4	9	8
5	8	9	4	3	2	6	7	1
6	3	4	7	5	1	2	8	9
7	1	2	8	9	6	3	4	5

208.

9	6	7	3	2	5	4	1	8
8	5	3	4	1	9	2	7	6
4	1	2	6	8	7	5	9	3
3	2	6	9	5	8	1	4	7
5	9	8	1	7	4	3	6	2
7	4	1	2	6	3	8	5	9
6	3	5	8	9	1	7	2	4
2	7	4	5	3	6	9	8	1
1	8	9	7	4	2	6	3	5

Cage sums: 22, 13, 16, 7, 3, 14, 8, 7, 16, 9, 8, 20, 9, 4, 7, 5, 11, 23, 7, 14, 7, 11, 11, 11, 8, 14, 20, 14, 9, 8, 18, 13, 15, 13, 10

209. There are 14 triangles. A good way to approach a puzzle like this is to letter each vertex and then go through the alphabet in order, three lengths at a time—for the three points that determine a triangle.

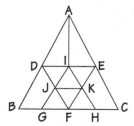

Δ ABC	Δ CEF	Δ FHK
Δ ADE	Δ DEF	Δ FJK
Δ ADI	Δ DIJ	Δ GHI
Δ AEI	Δ EIK	Δ IJK
Δ BDF	Δ FGJ	

210. There are 17 triangles. The triangles are:

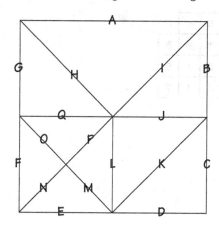

1.	E-M-N	10.	A-I-H
2.	F-O-N	11.	B-I-J
3.	O-P-Q	12.	J-K-L
4.	L-M-P	13.	C-D-K
5.	F-E-MO	14.	A-IPN-FG
6.	Q-L-MO	15.	BC-ED-NPI
7.	F-Q-NP	16.	NP-H-FG
8.	E-L-NP	17.	JQ-OM-K
9.	G-H-Q		

211.

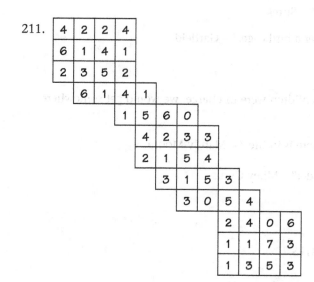

212.

```
3 1 7 4
6 4 1 3 1
2 7 1 2 3
4 3 4 1 3 0
    2 0 5 6 2
    5 1 7 2
      2 1 7 4 1
      0 2 3 7 3
      1 2 4 0 8
        3 5 2 5
        1 2 1 2 9
          1 3 2 5 4
          5 0 1 6 3
            3 6 1 5
            1 3 4 7
```

213. 1. 125 blocks total

2. 113 blocks total

3. 110 blocks total

4. 74 blocks total

214. a. "Better out than in, I always say." – Shrek

b. "Never leave your food dish under a bird cage." – Garfield

215. Son of a Gun

216. "I think that maybe if women and children were in charge, we would get somewhere."
– James Thurber

217. "Hope is the most exciting thing there is in life." – Mandy Moore

218. "Pink isn't just a color, it's an attitude!" – Miley Cyrus

219. To instruct others: TEACH

A melody of sound: MUSIC

How a speaker makes sound: VIBRATES

To choke: STRANGLE

A wicked hag: WITCH

To be confused or bewildered: BOGGLE

To achieve a great goal, one must begin with a small achievement.

220. Code Key: letters are assigned in opposite order (i.e., A = Z to Z = A)

1. Zinc (Zn)

2. Helium (He)

3. Nickel (Ni)

4. Hydrogen (H)

5. Carbon (C)

6. Silver (Ag)

7. Chlorine (Cl)

8. Copper (Cu)

9. Gold (Au)

10. Oxygen (O)

221. The figure opposite W is 7. When the W is oriented like it is here, the 7 looks like:

222.

		→ 14
7	8	→ 15
6	3	→ 9

↓ 13   ↓ 11

223. Cut the 2 × 12 rectangle into two pieces as shown . . .

. . . and reassemble like this.

224.	CUBE	C
	OBTUSE	O
	NUMBER	N
	GRAPH	G
	REMAINDER	R
	UNIT	U
	EXPONENT	E
	NORMAL	N
	TANGENT	T

225.	PRIME	P
	OVAL	O
	LINE	L
	YARDSTICK	Y
	GEOMETRY	G
	OUTPUT	O
	NUMBER	N

226.	SQUARE	S
	UNION	U
	BISECT	B
	TANGENT	T
	REAL	R
	ABSOLUTE	A
	CONSTANT	C
	TOTAL	T
	ISOSCELES	I
	OBLONG	O
	NATURAL	N

227.	VERTICAL	V
	OCTET	O
	LEVEL	L
	UNIT	U
	MATRIX	M
	EQUATION	E

228. Calm Down and Cheer Up

229. No one said these had to be straight lines. He swung an arc three times that divided the land properly.

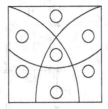

230. 16.

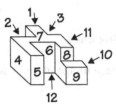

231. Start

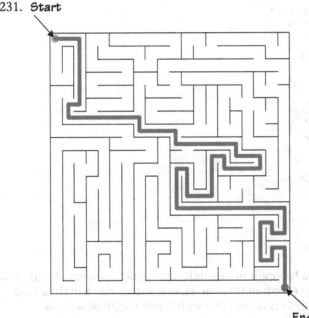

End

232. Here's one way.

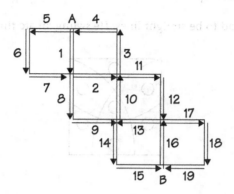

233. Illustration 5 can't be drawn without retracing or intersecting lines.

234. c. These are the consecutive capital letters and their reverse image placed back to back starting with the letter B.

235. Balanced Diet

236. Here's one way. Can you find another?

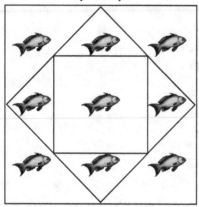

237. Summer is The Best Time of Year

238. There are 41 squares: 24 with a side length of one unit (the smallest square), 8 with a side length of two units, 6 with a side length of three units, one with a side length of four units, one with a side length of five units, and one with a side length of six units.

239. P is the last letter, and the sentence reads, "PUZZLES ARE A TRIP."

240. c. X cannot fit inside Y. X is the largest box because W fits inside it, Z is the same size as W, and Y fits inside Z.

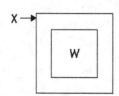

241.

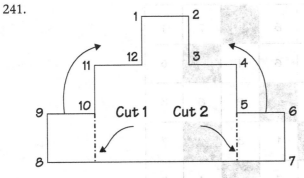

242. 33. Here are the counts for each block:

A.	B.	C.	D.	E.	F.	G.	H.	I.	J.	K.	L.	M.	N.
3	5	3	3	2	3	3	1	2	2	1	2	2	2

243.

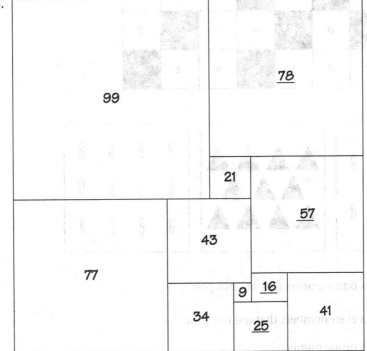

244. Extended Visit (OR A Long Visit)

245.

5	6	7	3	2	8	5	7	1
3	4	8	6	7	6	1	2	5
4	1	8	4	8	4	2	6	3
4	8	5	2	7	1	3	4	7
2	7	7	1	5	3	3	4	2
1	1	2	7	4	5	6	8	4
2	5	3	1	6	2	7	1	8
3	2	5	8	1	4	5	5	6
5	3	7	6	1	3	4	2	3

246.

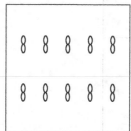 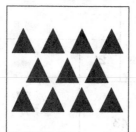 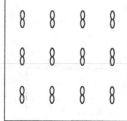

 Represents odd numbers that are not prime

 Represents even numbers that are not prime

Represents prime numbers

Note: Any reasonable arrangement of the symbols in the Answer section is acceptable.

247. B.

248. The letters of the words on the left are in alphabetical order; the letters in the words on the right are in reverse alphabetical order.

249. Six toothpicks have to be removed. Here's one way to solve this. Can you find others?

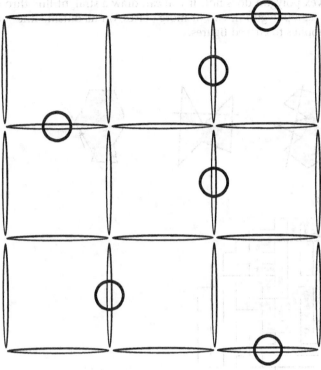

250. Eight sheets of paper are needed.

```
 ┌─────┬─────┬─────────┐
 │     │     │    6    │
 │  1  │  2  ├─────────┤
 │     │     │         │
 ├─────┴─────┤    7    │
 │     3     │         │
 ├───────────┼─────────┤
 │           │         │
 │           │ This is the size
 │     4     │ of the full sheet.
 │           │         │
 │           │    8    │
 ├───────────┤         │
 │     5     │         │
 └───────────┴─────────┘
```

251. The odd image out is C—the only illustration that has concave parts. The other illustrations are all convex illustrations. A concave polygon has one or more interior angles greater than 180°. A convex polygon does not. If you can draw a straight line through an illustration and intersect a polygon in two or more places, on its outside lines, it is concave. The same rule applies to curved figures.

Concave polygons:

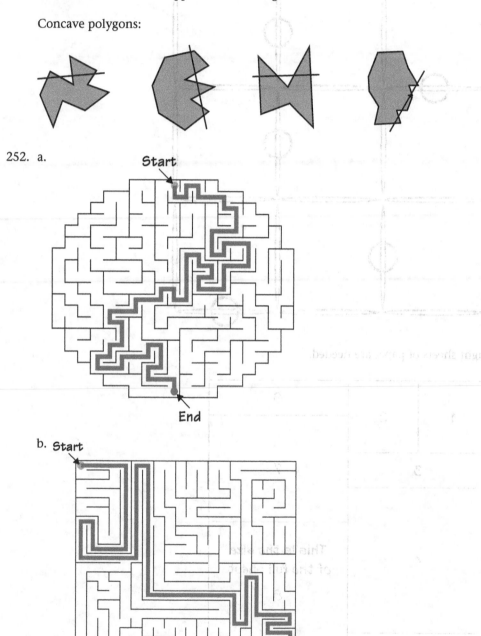

252. a.

Start

End

b. Start

End

253. Half-Time Performance

254. Figure 6. The other five figures each have two points of intersection. Figure 6 has four points of intersection.

255. B can be completed but will end at a different point than the starting point.

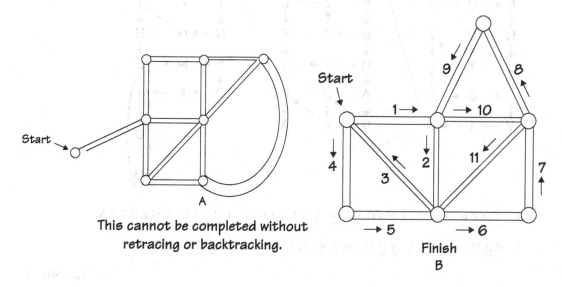

This cannot be completed without retracing or backtracking.

256. Stepfather

257. c. 3. Try different intervals, and make a chart of your findings.

258. Here's another answer. Did you find others?

	O		O		
O				O	
		O			O
	O		O		
		O			O
O				O	

Note: This solution is not symmetrical, so you could count its mirror image as an additional solution.

259.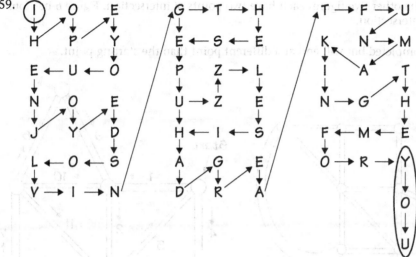

I HOPE YOU ENJOYED SOLVING THESE PUZZLES.

I HAD GREAT FUN MAKING THEM FOR YOU.

— Terry Stickels

Other

260. Sara is the oldest, then Gina, then Eileen.

261. The letters on the left are the left-most letters on a standard keyboard; the letters on the right, the right-most letters.

262.

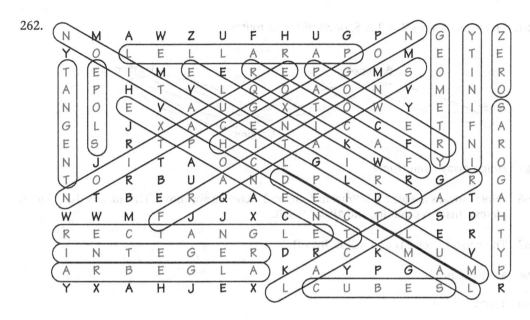

(Over, Down, Direction)

ALGEBRA (7,14,W)	CUBES (9,15,E)	DECIMAL (8,9,SE)
EQUATION (8,3,SW)	EXPONENT (3,5,SE)	FACTOR (12,8,NW)
FRACTIONS (4,11,NE)	GEOMETRY (13,1,S)	HEXAGON (6,7,NE)
INFINITY (14,8,N)	INTEGER (1,13,E)	MULTIPLICATION (14,14,NW)
PARALLEL (10,2,W)	POWER (9,3,SE)	PYTHAGORAS (15,14,N)
RADICAL (14,9,SW)	RECTANGLE (1,12,E)	SLOPE (2,7,N)
TANGENT (1,3,S)	TRIANGLE (12,10,NW)	ZERO (15,1,S)

263. 1, 4, 8

1, 5, 7

2, 3, 8

2, 4, 7

2, 5, 6

3, 4, 6

264. The jogger knows $2 \times 3 \times 5$, or 30 different routes.

265. Home Away From Home

266. "Nola" means apples. "Roz" means four, and "Kar" means move. This means that "Kir" is plates, "Insa" is carefully, and "Pala" is sell.

267. The correct choice is d. B will fit into D.

268. Mathematically

269. 17071.

270. 82,888.

271. Tongue in Cheek

272. b. path or way. Odometer has the prefix "odo."

273.

rate	art	tea
ate	ear	rut
a	tear	err
quart	quatre	rat
at	true	are
rare	truer	tare
tar	eat	rue

274. T, V, and X. The capital letters in group a are created with four straight strokes. The capital letters in group b are created with three straight strokes. The capital letters in group c are created with two straight strokes.

275. There are 99 palindromic numbers between 10 and 1000. There are 109 between 10 and 2000.

11	22	33	44	55	66	77	88	99	101	111
121	131	141	151	161	171	181	191	202	212	222
232	242	252	262	272	282	292	303	313	323	333
343	353	363	373	383	393	404	414	424	434	444
454	464	474	484	494	505	515	525	535	545	555
565	575	585	595	606	616	626	636	646	656	666
676	686	696	707	717	727	737	747	757	767	777
787	797	808	818	828	838	848	858	868	878	888
898	909	919	929	939	949	959	969	979	989	999
1001	1111	1221	1331	1441	1551	1661	1771	1881	1991	

276. Lon has four total cards—one each of the Yankees, Red Sox, Dodgers, and White Sox.

277. Barbara—Fly

Bianca—Reelo

Patty—Wheezie

Tina—Blaze

From c, we know Barbara can't be Blaze or Reelo. We also know from a that Patty can't be Reelo or Fly, and from c that she cannot be Blaze. That means she must be Wheezie and Barbara must be Fly. From b, Reelo is a faster swimmer than Tina, which means Reelo must be Bianca, leaving Tina to be Blaze.

	Reelo	Fly	Wheezie	Blaze
Barbara	O	X	O	O
Bianca	X	O	O	O
Patty	O	O	X	O
Tina	O	O	O	X

278. There are 15 different possible combinations:

Cherry ___ Vanilla	1	Vanilla ___ Chocolate	6	Chocolate ___ Mint	10
Chocolate	2	Mint	7	Mocha	11
Mint	3	Mocha	8	Strawberry	12
Mocha	4	Strawberry	9		
Strawberry	5				

Mint ___ Mocha	13	Mocha ___ Strawberry	15
Strawberry	14		

279. Scratch My Back

280. Fill 7-gallon container and pour it into 9-gallon container.

Fill 7-gallon container again and pour 2 more gallons into 9-gallon container.

Dump 9-gallon container and pour the 5 gallons from the 7-gallon container into the 9-gallon container.

Fill 7-gallon container.

Fill 9-gallon container from the 7-gallon container, leaving 3 gallons in the 7-gallon container.

Dump the 9 gallons and put the 3 gallons from the 7-gallon container into the 9-gallon container.

Fill the 7-gallon container one last time and fill the 9-gallon container with 6 more gallons from the 7-gallon container (filling it up), and leaving 1 gallon in the 7-gallon container.

281. c. Moles are the slowest and largest.

Speed	Size
Ants	Moles
Snails	Snails
Moles	Ants

282. 9006 is the next year after 6969. The year before 1691 is 1111.

283. b. Molly's father

284. The correct grid is A. The square starting in the upper left-hand corner of grid 1 moves downward one square in each grid. The square in the upper right-hand corner of grid 1 moves diagonally one square at a time. In the last 4 × 4 grid, they both meet in the same box, so there is only one square.

285. Drawer. The other words are palindromes. They read the same way both forward and backward. "Drawer" is "Reward" when read backward. Words of this kind are called recurrent palindromes.

286. e. Magnet. The others function only when they are in an upright position.

Part IV. Algebra, Statistics, and Probability

287. Bob is age 30, and Bill is 20. The sum of their ages ten years ago would be: $x + y = 30$. But Bob was twice as old, so:

$$2x + x = 30$$

$$3x = 30$$

$$x = 10 \rightarrow 10 \text{ years later, Bill is 20 and Bob is 30.}$$

288. I = 4P. Most people write the reverse.

289. The number is one.

Set up the puzzle like this:

7 times a number = 3 times that number plus 4

Let x = the mystery number

$$7x = 3x + 4$$

$$4x = 4$$

$$x = 1$$

290. 3. Let's say the number is x.

$$1\frac{1}{2} + x = 1\frac{1}{2} \times x$$

$$1\frac{1}{2} + x = 1\frac{1}{2}x$$

$$1\frac{1}{2} = 1\frac{1}{2}x - x \qquad \text{(moving the x from the left side to the right side)}$$

$$1\frac{1}{2} = 1\frac{1}{2}x$$

$$\frac{3}{2} = \frac{1}{2}x$$

$$x = 3$$

291. You are AHEAD of the Game

292. 27. Let's call the first digit X and the second digit Y. Then you have:

$$10X + Y = 3X + 3Y$$

$$7X = 2Y$$

Therefore, X = 2 and Y = 7

293. A = 3. Even if AB × C were its potentially highest value (97 × 5 = 485), A would be 4 at its highest value. Since we are considering only odd numbers, A must be 1 or 3. The only case where A could equal 1 is if B and C were 7 and 3 (or 3 and 7), respectively. Neither will result in 3 as an answer. Therefore, A = 3.

$$\begin{array}{r} 37 \\ \times\, 9 \\ \hline 333 \end{array}$$

294. c. $36.00. Ed paid $60.00 and sold it to Fred for 20 percent off of $60.00, or $48.00 (60 × .20 = 12; 60 − 12 = 48). Fred then sold it to Ned for 25 percent off $48.00, or $48.00 − $12.00 = $36.00 (25 percent of 48 is 12; 48 − 12 = 36). The total percentage discount from Ed to Ned was 40 percent or $24.00. $\frac{24}{60} = \frac{2}{5}$ = 40 percent.

295. A = T − 8

Since A + B = Z, then A + B + P = T. We know B + P = 8, so A + 8 = T. Therefore, A = T − 8.

296. c. 3P + 5280Q

There are 3 feet in a yard, so multiply 3 times the number of yards (P) to find the total feet in yards.

There are 5,280 feet in a mile, so multiply 5,280 by the number of miles to get the total feet in a mile.

Then add the two results to find the total number of feet in both.

297. 26. $\diamondsuit$ = 2

★ = 6

298. It is also 64.

$$4^n = 64$$

$$n = 3$$

2^{n+3}, where $n = 3$

$$2^6 = 64$$

299. a. 9

$$\triangle_6 = 6^2 - 1 = 35$$

$$\boxed{5} = 5^2 + 1 = 26$$

$$35 - 26 = 9$$

b. 73

$$\triangle_8 = 8^2 - 1 = 63$$

$$\boxed{3} = 3^2 + 1 = 10$$

$$63 + 10 = 73$$

300. c. A square number.

$$\frac{P}{Q} = \frac{4}{5}$$

Cross multiplying, $Q \times 4 = 5 \times P$.

Since the two above quantities are equal, if they are multiplied, it will result in a square number.

301. b. $\frac{M}{H} \times Y$. $\frac{M}{H}$ = Miles per hour. Y is the number of hours you are traveling at $\frac{M}{H}$.

So, $Y \times \frac{M}{H}$ will give you miles.

$$(Y)(\text{hours}) \times \frac{(M)\text{miles}}{(H)(\text{hours})} = \text{miles}$$

302. No Horsing Around

303. 250. 40% of 50 is 20.

Then, 8% of some number is equal to 20.

$$.08x = 20$$
$$8x = 2,000$$
$$x = 250$$

304. 3 points. Penny had three arrows in "A" and one in "B" = 14 pts. Molly had one arrow in "A" and one in "B" = 11 pts.

The only difference between the two is that Penny had one more arrow and it landed in A. She also had 3 more points than Molly so that one extra arrow in A must be worth 3 points!

305. What does a duck do when it flies upside down?

It quacks up.

306. To solve the equations:

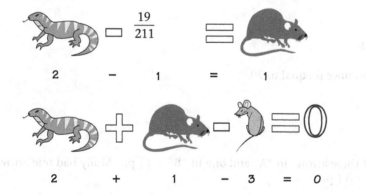

 = 3   = 2   = 1

Therefore:

3 − 2 = 1

2 + 1 − 2 = 1

2 + 2 − 1 = 3

$$2 - 1 = \frac{19}{211}$$

2 + 1 − 3 = 0

307. 1.
$$\begin{array}{r} 533 \\ -\ 412 \\ \hline 121 \end{array}$$

2.
$$\begin{array}{r} 336 \\ 312 \\ +\ \ \ 3 \\ \hline 651 \end{array}$$

3.
$$\begin{array}{r} 541 \\ -\ 325 \\ \hline 216 \end{array}$$

4.
$$\begin{array}{r} 162 \\ 43 \\ 26 \\ +\ 34 \\ \hline 265 \end{array}$$

308. d. D = 0.

309. 24. 16 of the players are either outfielders or infielders, leaving 8 players to be pitchers. $\frac{8}{24} = \frac{1}{3}$. Solving this with algebra:

$$16 + \frac{1}{3}x = x$$
$$48 + x = 3x$$
$$2x = 48$$
$$x = 24$$

310.
$$\begin{array}{r} 372 \\ 372 \\ 372 \\ +372 \\ \hline 1488 \end{array}$$

$$\begin{array}{r} \text{FUN} \\ \text{FUN} \\ \text{FUN} \\ +\ \text{FUN} \\ \hline \text{BALL} \end{array}$$

A = 4
B = 1
F = 3
L = 8
N = 2
U = 7

311. a. Working from inside the parentheses first, $5 \triangle 7 = \frac{5+7}{3}$. So $5 \triangle \left(\frac{5+7}{3} \right) = 5 \triangle 4$.

Now, $5 \triangle 4 = \frac{5+4}{3} = 3$

b.
$$\frac{6 \triangle 3 = \frac{6+3}{3} = 3}{7 \triangle 8 = \frac{7+8}{3} = 5} = \frac{3}{5}$$

312. The largest value of $\frac{x-y}{x+y}$ is $\frac{99}{101} \ldots \left(\frac{100-1}{100+1} \right)$

The largest value of $\frac{x+y}{x-y}$ is $\frac{199}{1}$ or $199 \ldots \left(\frac{100+99}{100-99} \right)$

313. Kick the Bucket

314. The ratio is 3 to 2. You can set up this puzzle like an algebra problem without having to solve for x or y. Let x be the number of men and y be the number of women. So the total of both men's and women's ages is $36 (x + y)$. But that number is equal to $45 \times x + 30 \times y$.

$$36(x + y) = 45x + 30y$$

$$36x + 36y = 45x + 30y$$

$$6y = 9x$$

$$y = \frac{9}{6}x$$

$$y = \frac{3}{2}x$$

So the ratio of women to men is 3 to 2.

315. $Z = 8$

$$\begin{array}{r} YY \\ +YY \\ \hline XYZ \end{array} \qquad \begin{array}{r} 99 \\ +99 \\ \hline 198 \end{array}$$

You know that Y has to be greater than 5 because you carry a 1 to make X. You also see where $\begin{array}{r} Y \\ +Y \\ \hline Y \end{array}$, which means that a 9 is the only number that works. Carry the "1" over from the first column: $\begin{array}{r} {}^{1}9 \\ +9 \\ \hline 19 \end{array}$

316. $\bullet = 3$

$\Delta = 7$

$* = 4$

Look at the 2 triangles on the left side of scale b. They are equal to $* * \bullet \bullet$. Visually move $* * \bullet \bullet$ to replace the two triangles on the right-hand side of scale a. Now you have $* * * * * = * * \bullet \bullet \bullet \bullet$ in scale a. This reduces to $* * * = \bullet \bullet \bullet \bullet$. So we can tell that a $*$ is equal to 4. Now it is easy to see that the $\Delta = 7$.

317. 12 and 14.

With a difference of 2 and 168 ending in an even number, you know two things: The numbers are even consecutive numbers, one of which is close to 12 ($12^2 = 144$). It doesn't take long to find 14×12 if you solve it using algebra:

1. $x - y = 2$

2. $xy = 168$

3. $x = y + 2$

4. $(y + 2)y = 168$

5. $y^2 + 2y - 168 = 0$

6. $(y + 14)(y - 12) = 0$

7. $y = \underline{12}$

8. $x = y + 2 = \underline{14}$

318. $\frac{7}{10}$. There are several ways to solve this. One way is to see that 2 goes into 7 3.5 times

and $3.5 \times \frac{1}{5} = \frac{7}{10}$. Another way is to set up a proportion:

$$\frac{2}{\frac{1}{5}} = \frac{7}{x}$$

$$2x = \frac{7}{5}$$

$$x = \frac{7}{10}$$

319. b. 35x. If she has 2 nickels, she has 6 dimes for a total of $.70. If she has 3 nickels, she has 9 dimes, for a total of $1.05 ($.35 × 3 = $1.05).

320.

321. A = 7   D = 5   H = 0   I = 8   M = 9   P = 3   R = 2   S = 1   T = 4   U = 6

```
     755
      84
      63
    9740
  +   81
   10723
```

322. $1\frac{2}{3}$ pounds.

Let x = the whole carton.

$x = \frac{5}{8}x + \frac{5}{8}$ pound

Multiply both sides by 8.

$8x = 5x + 5$ pounds

$3x = 5$ pounds

$x = 1\frac{2}{3}$ pounds

323. 21 pieces. He just might pull four pieces of each of the five colors—and then the next piece he pulls would assure him of five pieces of one of the colors.

324. $\dfrac{6}{9} \times \dfrac{5}{8} = \dfrac{30}{72} = \dfrac{15}{36} = \dfrac{5}{12}$

Your chances are 5 in 12. There are 6 black cubes in a total of 9 cubes. So on your first pull your chances are 6 in 9 of drawing a black cube. Your next selection will be from 8 cubes, 5 of which are black. You then multiply those two chances together: $\dfrac{6}{9} \times \dfrac{5}{8} = \dfrac{5}{12}$.

325. 1 in 4. The area of the larger circle is $A = \pi r^2 = \pi 3^2$.

The area of the smaller circle is $A = \pi r^2 = \pi 1.5^2$.

$$3^2 = 9$$

$$1.5^2 = 2.25 \qquad \dfrac{2.25}{9.00} = \dfrac{1}{4}$$

326. Here's one way to look at this:

40 added to $\dfrac{1}{2}$ of some number (call it x) = 3 times that number ($3x$).

$$40 + \dfrac{1}{2}x = 3x$$

$$2\dfrac{1}{2}x = 40$$

$$\dfrac{5}{2}x = 40$$

$$5x = 80$$

$$\underline{x = 16}$$

327. 25 gallons.

$$\dfrac{25 + x}{100 + x} = \dfrac{.4}{1}$$

$$25 + x = .4(100 + x)$$

$$25 + x = 40 + .4x$$

$$.6x = 15$$

$$x = 15 \div .6$$

$$\underline{x = 25 \text{ gallons}}$$

So the next container would have 25 gallons of the chemical and a total of 125 gallons.

Additional Math Resources from Terry Stickels

MATH PUZZLES AND BRAINTEASERS
GRADES 3-5

Over 300 Puzzles that Teach Math and Problem-Solving Skills

Terry Stickels

ISBN: 978-0-470-22719-0
Paperback | 248 pp.

Math Puzzles and Brainteasers, Grades 3-5 contains over 300 reproducible puzzles and brain games that teach math skills, problem solving, and critical thinking—all organized into groups that correspond to concepts recommended for learning by Standards from the National Council for Teachers of Mathematics, including:

- Numbers and Operations
- Geometry and Measurement
- Mathematical Reasoning
- Algebra, Statistics, and Probability

This resource also includes a wide range of popular puzzle types like sudoku, kokuru, and frame games as well as math, visual and logic puzzles, word problems, and other brain-twisters that work to teach mathematical fundamentals while engaging students' interests and attention.

Praise for *Math Puzzles and Brainteasers, Grades 3-5*

"This is a rich set of diverse mathematical problems that can enrich a math class, stimulate children to play with mathematical ideas, or give gifted children a chance to solve interesting problems that lie beyond the limits of most school mathematical curricula."
—**Michael Schiro,** associate professor, Mathematics Education, Boston College

"What an extraordinary variety of intriguing and challenging mental games! It is exactly what is needed to help children develop thinking and problem solving skills."
—**Pat Battaglia,** author of many puzzle books and the syndicated column If You're So Smart…

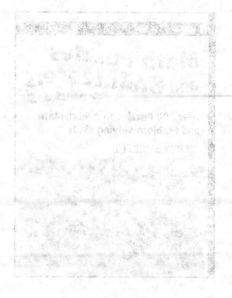